Cells and Stem Cells
The Myth of Life Sciences

Cells and Stem Cells
The Myth of Life Sciences

Dianliang Wang
China Association for Science and Technology, China

Haijia Chen
Guangzhou Saliai Stem Cell Science and Technology Co., Ltd., China

 Chemical Industry Press Co., Ltd.    **World Scientific**

Published by

World Scientific Publishing Co. Pte. Ltd.

5 Toh Tuck Link, Singapore 596224

USA office: 27 Warren Street, Suite 401-402, Hackensack, NJ 07601

UK office: 57 Shelton Street, Covent Garden, London WC2H 9HE

and

Chemical Industry Press Co., Ltd.
No. 13, Qingnianhu South Street
Dongcheng District, Beijing 100011
P. R. China

Library of Congress Control Number: 2021946323

British Library Cataloguing-in-Publication Data
A catalogue record for this book is available from the British Library.

CELLS AND STEM CELLS
The Myth of Life Sciences

ISBN 978-981-123-877-2 (hardcover)
ISBN 978-981-123-978-6 (paperback)
ISBN 978-981-123-878-9 (ebook for institutions)
ISBN 978-981-123-879-6 (ebook for individuals)

For any available supplementary material, please visit
https://www.worldscientific.com/worldscibooks/10.1142/12329#t=suppl

Typeset by Stallion Press
Email: enquiries@stallionpress.com

Foreword

The sciences that are closely related to human survival and self-development are nothing but the life sciences. At present, a new round of global scientific and technological revolution and industrial transformation is emerging. Biochemistry, genetics, cell biology, and molecular biology are all developing rapidly. Major breakthroughs have continuously been made in gene sequencing, cell therapy, molecular breeding, protein engineering, and other biotechnologies, which provide strong support for human beings to cope with challenges such as health, food and energy, and economic and social development, and have already had profound impacts on economic and social developments and people's production and lives.

All these are inseparable from the understanding of life phenomena. In 1663, Robert Hooke discovered cells; in 1839, Jakob Schleiden and Theodor Schwann jointly put forward cell theory, pointing out that cells are the basic units of structure and function of organisms. Subsequently, biological research opened a colorful world to people, such as evolution theory, genetics and DNA double helix model, genetic engineering, human genome project, and more. Scientists have gradually uncovered the mysterious veil of life. Many young people have strong interests in the phenomena of life, and thus begin their journey of scientific explorations, witnessing the amazing miracles of lives.

The study of life sciences usually starts with cells. Biology textbooks start with cell morphology and go all the way to individuals and communities, genetics and evolution, environment and ecology. Whether it is a lower organism or a higher organism including human beings, the basic structural unit is a cell, and life activities are derived from cells. Therefore, to understand life, we must first understand cells, understand the various life activities in the cells. *Cells and Stem Cells: The Myth of Life Sciences* edited and largely written by Professor Dianliang Wang describes the colorful scenes of the cell life world for us.

Professor Dianliang Wang has been engaging in research on stem cells, tissue engineering, and regenerative medicine for decades, and has published more than 8 million words of academic monographs. Besides the intense scientific research, he spent a lot of effort on science popularization and science fiction creation. *Cells and Stem Cells: The Myth of Life Sciences* systematically introduces the knowledge, technologies, and theories of cells and stem cells, and various products of cells and stem cells that have brought changes to our lives. This book is scientific, interesting, illustrated, informative, and also easily understandable. It is a rare excellent popular science book indeed. The ultimate mission of science research is to construct a better future for human beings, and the yearning for the better future can start with excellent popular science books.

Hongxiang Zhang
Director of the Science Popularization Working Committee
of the Chinese Society of Biotechnology
Beijing, China
November 26, 2020

Preface

The History of Human Understanding of Cells

About 4 billion years ago, lives were born on this beautiful light blue planet where we live. The lives that were just born were only some rather rudimentary aggregations of molecules, without cell structure. To this day, we still do not know what the original lives were, but we can infer from the existing biological knowledge that they could communicate with the external environment and roughly replicate themselves. With the evolution of biology, lives in the form of cells finally appeared in nature.

The oldest fossils on earth are found in Australia, called stromatolites. The organisms in the fossils are blue-green algae, which lived about 3.5 billion years ago. Unlike animal and plant cells, blue-green algae have no typical cell nuclei. They are similar to bacteria, also known as cyanobacteria. They are unicellular organisms.

These unicellular blue-green algae are all over the primitive oceans. They are spherical, tubular, and have complex cellular structures. Like the green plants that live on the earth today, blue-green algae can use the energy of the sun to convert carbon dioxide and water into organic compounds that they need and release oxygen. With the emergence of oxygen, the speed of biological evolution was greatly accelerated. At the same time, in order to better adapt to the living environment, a number of unicellular algae combined to form multicellular community organisms, just like today's gonium and pandorina, between unicellular organisms and multicellular organisms. As organisms continued to evolve, multicellular plants were born about three billion years ago. Then, multicellular animals came into the world. The lives in the sea began to move to land, and the lives on land continued to evolve. Then, only a few million years ago, the original human beings were finally born. Human beings are composed of about 100 trillion cells, and the DNA molecules in each cell contain all the genetic information of the human body.

With the birth of human beings, the dawn of civilization began to appear. However, for a long time after the birth of human beings, the mystery of life, the cell, was not discovered. This is mainly because the vast majority of cells are too small, below 0.03 millimeter, which is below the smallest size of 0.1 millimeter that the human naked eye can directly see.

The first person who had actually observed living cells was Antonie van Leeuwenhoek, a Dutch scientist. In 1677, he used a self-made microscope to observe protozoa, the sperms of human and mammals in pond water, and then saw the salmon red blood cell nucleus. In 1683, bacteria were found in tartar. Leeuwenhoek made a detailed record of the observations and then communicated them to the Royal Society of England. The Royal Society highly praised his achievements.

The discovery of living cells promoted the development of cell biology. In 1938, the German botanist Matthias Schleiden discovered that all plants are composed of cells. One year later, the German zoologist Theodor Schwann discovered that all animals are also composed of cells. Matthias Schleiden and Theodor Schwann co-founded the cell theory, which includes three parts: First, a cell is the smallest structural unit of multicellular organisms. For unicellular organisms, a cell is a biological individual. Second, each cell of multicellular organisms performs specific functions. Third, cells can only be derived from cell divisions. This theory makes clear the unity between animals and plants and is known as one of the three major discoveries of natural sciences in the 19th century.

After the establishment of cell theory, many scientists turned their attention to the contents of cells and discovered the living substance, protoplasm, in cells. By using the fixed staining technique, organelles such as centrosome, Golgi apparatus, and mitochondrion were discovered. Meanwhile, great progress was made in the study of cell division and chromosome. Soon afterward, people began to use cells, transform cells, produce valuable industrial and agricultural products, create new varieties of animals and plants, and serve human life and health.

In 1907, cell culture technology was established, which laid the foundation for the use of cells. In 1958, Japanese scientist Yoshio Okada discovered that Sendai viruses inactivated by ultraviolet radiation could cause the tumor cells of Ehrlich ascites to fuse with each other. In 1965, Harris induced fusion of different animal somatic cells. Unexpectedly, the hybrid cells survived. This is a new type of engineered cells, but it has no practical use. The year 1975 is memorable in cell history. This year, immunologists Kohler and Milstein used Sendai viruses to induce the fusion of mouse spleen cells immunized with sheep red blood cells and mouse myeloma cells to select a kind of hybrid cell that is capable of secreting monoclonal antibodies.

Today, monoclonal antibodies are widely used in disease diagnosis and cancer treatment, and have the reputation of being "biological missiles".

Through plant cell culture, a large number of valuable flowers have been produced, such as clivia, hyacinth, and carnation, as well as precious Chinese herbal medicines, such as ginseng, angelica, and pseudo-ginseng. Animal cell mass culture is also very attractive: One is the production of vaccines and the other is the production of high-value protein drugs for the treatment of tumors, cardiovascular diseases, and so on.

The research on animal cloning has made cell technology a high technology, attracting worldwide attention. In 1952, American scientists used a tadpole cell to create an exact replica of the original, and the tadpole became the first cloned animal in the world. In 1996, Dolly, the first adult somatic cell cloned sheep in the world, was born at the Roslin Institute in Edinburgh, Scotland. It was proved for the first time that animal somatic cells and plant cells have the same genetic totipotency, breaking the traditional scientific concept and making a sensation in the world. In 1998, scientists at the University of Hawaii cloned more than 50 mice from adult somatic cells and began to clone animals in batches. In 2008, the US Food and Drug Administration announced that it approved the marketing of dairy and meat products from cloned animals, and claimed that these controversial foods could be eaten as safely as normal animal products. In 1999, stem cell research was promoted as one of the ten most important scientific research fields in the 21st century by the famous American journal *Science*, and ranked first, ahead of the "human genome project". In 2000, stem cell research was once again selected as one of the top ten scientific and technological achievements of that year by the journal *Science*. Since 2011, South Korea, the United States, Canada, and other countries have approved new stem cell drugs, enabling effective treatment of some difficult and complicated diseases. In 2012, China carried out the standardized management of stem cell therapy, and immune cell therapy achieved unprecedented development. In 2013, the American journal of *Science* listed tumor immunotherapy as one of the top ten scientific breakthroughs of that year. In February 2015, the European Medicines Agency approved the clinical application of human autologous corneal epithelial cell drug Holoclar containing stem cells from Italy's Chiesi Farmaceutici S.p.A for the treatment of moderate to severe limbal stem cell deficiency caused by physical or chemical burning in adult patients. In December 2015, China Boyalife Stem Cell Group Co., Ltd. collaborated with the Sooam Biotech Research Foundation of South Korea to build the world's largest cloning factory in Tianjin, China. It planned to produce 1 million cloned cattle each year, and will clone dogs and even endangered species. All of these have epoch-making significance.

In recent years, cell transplantation therapy has attracted worldwide attention. On January 12, 2018, the Nanjing Gulou Hospital announced the first healthy baby born in the clinical research of stem cell treatment of premature ovarian failure in China. It was also the first healthy baby born in the clinical research of stem cell composite collagen scaffold material for the treatment of premature ovarian failure in the world, and made a major breakthrough in stem cell tissue engineering and regenerative medicine. In August 2020, the Drug Controller General of India approved the launch of an adult allogeneic bone marrow mesenchymal stem cell drug named Stempeucel in India for the treatment of critical limb ischemia due to Buerger's Disease and Atherosclerotic Peripheral

Arterial Disease. The product was developed by the partner of Cipla Limited, Stempeutics Research Pvt. Ltd.

Cells are magical and are the life cradles. The cell sciences and technologies have changed our lives and made our lives better.

Dianliang Wang
Beijing, China
November 28, 2020

Contents

Book Summary

This popular science book systematically introduces major scientific and technological achievements in the field of cells and stem cells, and the conveniences they bring to human life. It covers plant cloning, animal cloning, human cloning, biological missiles, biological drugs, immunocytotherapy, stem cell therapy, stem cell bank, 4D printing, 5D printing, CAR-T technology, and other frontier fields, which reflect the latest progresses and development trends of life sciences. The book is both interesting and rich in information, revealing the magic and mystery of life sciences.

About the Authors

Dianliang Wang, PhD, Professor, National Chief Scientist of the China Association for Science and Technology in the field of stem cell and tissue engineering. He is the executive director of the Chinese Society of Biotechnology and the deputy director of the Professional Committee on Stem Cell and Tissue Engineering, the executive director of the China Quality Association for Pharmaceuticals and the deputy director of the Stem Cell Regenerative Medicine and Precision Medicine Quality Committee, and the deputy director and secretary-general of the International Human Gene Variation Group Project China Expert Committee. He is the author of 16 scientific books and 7 science fictions. He is an adjunct professor at many universities in China. His main research directions include stem cell tissue engineering and regenerative medicine, biomedicine, and new biological materials.

Haijia Chen, PhD, Founder and Chairman of Guangzhou Saliai Stem Cell Science and Technology Co., Ltd.; Director of Guangdong Saliai Stem Cell Research Institute.

Acknowledgments

In the process of writing this book, I received vigorous support from some academicians and experts. They are Professor Sir Christopher Pissarides, who is the winner of the Nobel Prize in Economics in 2010 and an academician of the British Academy of Social Sciences; Professor Kwok Fai So, who is an Academician of the Chinese Academy of Sciences, Director of Department of Ophthalmology, State Key Lab of Brain and Cognitive Sciences, The University of Hong Kong; Professor Shixin Lu, who is an Academician of the Chinese Academy of Sciences, Researcher and former Director, Institute of Oncology, Chinese Academy of Medical Sciences; Professor Weidong Le, who is the Director of the Institute of Neurology of the Affiliated Sichuan People's Hospital of the University of Electronic Science and Technology, China; and Professor Yen Wei, who is the Director of the Center for Frontier Polymer Research of Tsinghua University, China.

Academician Yong Zhang, Professor Haijia Chen, and Director Peiduo Zheng generously provided some images. On the occasion of the publication of this book, I would like to express my sincere gratitude to all the academicians, experts, and friends for their sincere advice and support.

Without them, this book would not have been possible. Due to limited time and our limitation of knowledge and experience, this book may have some errors and omissions. I hope that the dear readers will not hesitate to share some comments and the errors will be revised at reprinting.

Introduction

Cells, we are no stranger. What are the functions of cells besides constituting the bodies of animals and plants? Maybe a lot of people don't know. Through the operation of cells that cannot be directly seen by the naked eye, many amazing miracles of life can be created.

Some cells are taken from a green leaf and cultured in a glass bottle for a period of time, and the whole plant can be regenerated. Two plants with far different genetic relationships can produce a "double floor crop" as beautiful as the "potato-tomato" through cell fusion, where its above-ground part bears tomatoes and the underground part grows potatoes.

Animal cells can also bring surprises. A new hybrid of carp-crucian can be developed by transferring the nucleus of carp into the cell of a crucian, which has the advantages of two kinds of fish. The human embryonic stem cells can be used to prepare the liver, kidney, heart, and lung for organ transplantation. The hybridoma cells produced by the fusion of tumor cells and lymphocytes can secrete monoclonal antibodies, which can be combined with anticancer drugs to make biological missiles, and they can accurately attack the "nests" of cancer cells. Dolly, the cloned sheep, once stirred up the world; when cloned animals enter the stage of massive production, would you dare eat the meat of cloned animals? Immune cell therapy has been used in the treatment of tumors. Is it still far to conquer cancers? A variety of new stem cell drugs have been approved by many countries. Can stem cells make people more beautiful, healthier, and long-lived?

...

All in all, cells and stem cells are changing our lives.

Chapter 1

The Magical World of Cells

1. The Simple Life Forms in Nature

In the rich and colorful Kingdom of life, the simplest "citizens" are nothing but viruses, viroids, and prions [1]. Some of them are just biological macromolecules. However, when they are parasitized in living cells, they can show life activities again, which is very magical. Viruses, viroids, and prions are dwarfs that cannot be seen with the naked eye in the kingdom of life, and it is not easy to find them.

At the end of the 19th century, when researchers were studying tobacco mosaic disease and bovine foot-and-mouth disease, they found that the pathogens of these diseases could pass through the porcelain filters that bacteria could not pass through. At that time, they called these pathogens filterable viruses or viruses to distinguish them from the pathogens of many other diseases, bacteria.

The size of viruses varies greatly, ranging from 10 to 30 nanometers (one nanometer equals one millionth of a millimeter). There are also various shapes, including cubic symmetry, spiral symmetry, and complex symmetry, but the composition is very simple. Many viruses only contain nucleic acids and proteins. Some viruses such as the influenza virus also carry the components of host cell membranes when they are released from host cells, so they contain a small amount of saccharide and fat substances. However, all viruses contain only one kind of nucleic acid. According to the types of the contained nucleic acids, viruses can be divided into DNA viruses and RNA viruses.

Viruses seem to be omnipresent "parasites", which not only dare to parasite on plants, animals, and humans but also refuses to let go of bacteria that can't be seen by the naked eye. According to the different parasitic objects, viruses can be divided into animal viruses, plant viruses, and bacterial viruses. Among them, bacterial viruses are also called bacteriophages.

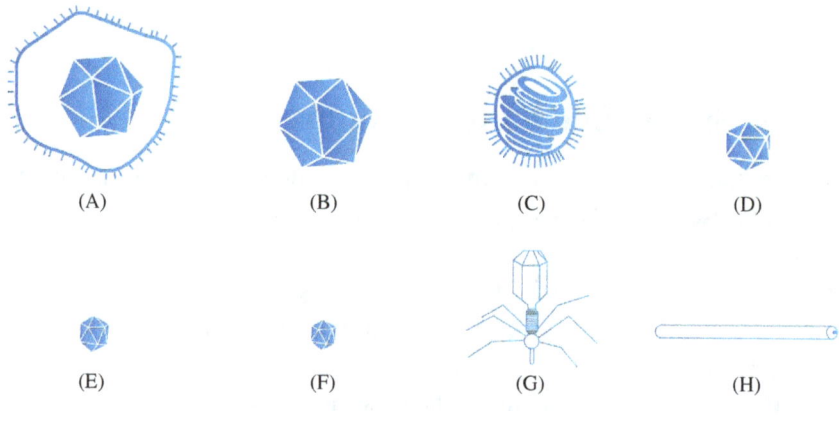

(A)   (B)   (C)   (D)

(E)   (F)   (G)   (H)

Different forms of viruses

(A) Herpesvirus; (B) mosquito virus; (C) influenza virus; (D) adenovirus; (E) polyomavirus; (F) poliovirus; (G) T-even phage; (H) tobacco mosaic virus.

Viruses are specifically parasitized in other organisms, and they need to use host materials to replicate themselves in the process of life, which will certainly cause great damage to the parasitized organisms. Most commonly, viruses can cause a variety of infectious diseases, some even horrible fatal epidemic infectious diseases, such as human avian influenza, Ebola hemorrhagic fever, AIDS, hepatitis A, epidemic encephalitis B, smallpox, measles, and polio. Many viruses can also cause human cancers, such as adenovirus and hepatitis B virus, which can make people suffer from cancers. In fact, most of the more than 300 viruses found so far can cause other organisms to get sick.

But, it is not that viruses have no use for humans at all. With the development of science and technology, various viruses are being used for the benefits of mankind. First of all, after special treatment, some viruses can be made into live attenuated vaccines, such as measles live attenuated vaccine, mumps live attenuated vaccine, and oral rabies live attenuated vaccine, to prevent various severe infectious diseases. Second, some viruses, such as insect baculovirus, can be used as vectors for gene engineering operations. Third, it is also very interesting that many insect viruses are specifically parasitized in some agricultural insect pests and spread severe pestilence among them. However, they are not toxic to other animals, plants, and humans. If these insect viruses are made into biological pesticides or insecticides after industrial production, and are sprayed in the forest or field, they can play dual effects of protecting the environment and eliminating pests, which is incomparable to highly toxic chemical pesticides. In China, this high technology has already been applied.

Viruses are simple, but viroids are much simpler. They are only ribonucleic acid (RNA) molecules that are 80 times smaller than known viruses. The molecular weight of these RNA molecules is 75,000–85,000 daltons. Like viruses, viroids cannot live on their own and must live in living cells. The results of parasitism often lead to plant diseases, for example, potato spindle tuber disease is the result of viroid parasitism.

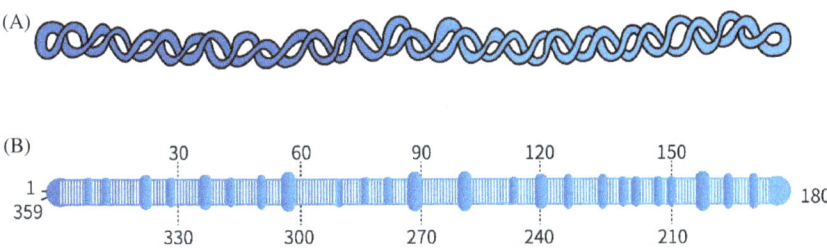

Potato spindle tuber viroid

(A) The virus particle is rod-shaped; (B) the RNA molecular structure of the virus.

Unlike viroids, prions contain only proteins [2], and were first discovered by American virologist Prusiner in 1982. These proteins have infectivity and are pathogenic factors of sheep pruritus, which is a degenerative disorder of the central nervous system. After suffering from the disease, the animals were restless, covered with scabies, with their hair flaking off, and skin damage. About half a year later, the animals were obviously weak and out of balance. At the later stage, the limbs became numb. Finally, the sick sheep died in agony.

At first, many scientists did not believe in prions, because according to the traditional biological law, only nucleic acids can replicate themselves. As living bodies, how can prions replicate? Can proteins replicate themselves? Scientists suspect that prions may contain nucleic acids, which are lost during the extraction of pathogens. However, with the in-depth study of prions, it was found that this mysterious pathogen does contain only protein, and this protein can replicate itself. Thus, Prusiner won the Nobel Prize for the discovery of and research on prions.

So far, the story of prions is not over. In April 1985, a kind of strange mad cow disease (bovine spongiform encephalopathy, BSE) appeared in Britain. After more than 10 years, the disease spread rapidly to many countries in the world. After a certain incubation period, the infected cattle became sick and finally died painfully. After dissecting the carcasses of the diseased cattle, it was found that a large number of nerve cells were lost and amyloid lesions appeared in the brain, which soon became a mass of paste. A large number of sick cattle had to be slaughtered mercilessly, but even then, the epidemic was difficult to control.

Mad cow disease caused panic in many countries. To make matters worse, many people contracted Creutzfeldt Jakob disease, which was similar to mad cow disease. Some people thought that it was caused by eating beef infected with mad cow disease. The media all over the world seized the opportunity and carried out continuous reports. For a time, mad cow disease became a hot topic on the streets. Many countries mentioned banning the import of beef, beef products, and feed from Europe, which made many European countries miserable and suffer huge economic losses.

How can a healthy cow get disease? According to the latest scientific research, prions are the main culprit of the mad cow disease epidemic.

According to scientists, proteins with exactly the same chemical composition as prions are found in normal brain tissues, but with different configurations. As for how prions replicate themselves, and how normal proteins transform into prions, this is intriguing, and is also a question that scientists are studying in depth.

In fact, there are still controversies about whether viruses, viroids, and prions are life, but their discoveries greatly shorten the distance between life and non-life, and also strengthen the understanding of cell life.

2. The Colorful World of Cells

The earth we live on is an authentic paradise of lives, where there are more than 100,000 species of microbes, more than 300,000 species of plants, and more than 1 million species of animals. Among such a wide range of organisms, the simplest living creature that can live independently is probably mycoplasma. It was discovered in cell-free medium at a time similar to the discovery of virus, which was then called pleuropneumonia-like microorganism. Later, dozens of such microbes were found in soil, sewage, and many animals and human beings.

From their appearances, mycoplasmas are very similar to rice glue balls, and the stuffings are wrapped in thin skins [3]. The sizes of different mycoplasmas vary greatly, usually in the range of 0.1–0.25 micron (one micron is one thousandth of a millimeter), and the volume of the smallest mycoplasma is only one thousandth of the size of ordinary bacteria. It can pass through a filter like viruses and can also grow in artificial media like bacteria, so it is a transitional organism between viruses and bacteria. The skins of mycoplasmas are similar to the general cell membranes, which are double layers and consist of phospholipids and proteins. The stuffings contain the substances using which mycoplasmas carry out life activities, such as DNA and RNA that store and transmit life information, as well as various enzymes involved in metabolisms.

Different from the parasitic lifestyle of viruses, mycoplasmas can absorb nutrients from artificial culture media and lead a completely free life, but they are still the pathogens of many diseases, and some can cause arthritis in pigs and others can cause pneumoniae in humans.

Bacteria are much more complex than mycoplasmas, and may mainly be spherical, rod-shaped, and spiral. Their sizes are generally less than 1 micron. From outside to inside, a bacterium is divided into a cell wall, cell membrane, cytoplasm, and nucleoid. The so-called nucleoid is not the real nucleus, but a mass of genetic materials dispersed in the cytoplasm. For this reason, bacteria are also called prokaryotes.

The mode of reproduction of bacteria is relatively simple. For the vast majority of bacteria, the genetic materials are first replicated before reproduction, and then divided into two parts from the middle of each bacterium. A few bacteria propagate by spores or budding. Bacteria are highly efficient in reproduction. As for *Escherichia colis* [4], which are widely distributed in waters and in the intestines of animals and human beings, they reproduce one generation about every 20 minutes, making the bacteria almost ubiquitous on earth. Bacteria also have an expertise. When the living environment becomes bad, they will turn into spores, which can resist the adverse environment. When the conditions are suitable, the spores will germinate like seeds and can be grown into new bacteria. Some bacteria are covered with cilia, and others have long flagella-like whips. These things are not just for decoration, they are the locomotive organs of bacteria. For pathogenic bacteria, the cilia on the surfaces of bacteria are also conducive to attachment to animal and plant cells. Many kinds of bacteria feed on sulfur, iron ore, and oil, which are not nutritious to human beings. It is incredible.

In people's minds, bacteria are the things to avoid. In daily life, the thought of bacteria is always associated with infection, fever, inflammation, suppuration, and even terrible diseases such as tetanus, gonorrhea, and syphilis. In fact, not all bacteria are harmful to human beings. *Escherichia colis*, which are parasitic in human and animal intestines, can help digestion and produce beneficial vitamins. In modern bioengineering, *Escherichia colis* are often used to transfer drug genes into engineered bacteria to produce protein drugs, such as interleukin, interferon, and erythropoietin. Other bacteria are used to metallurgy and clean up oil pollution at sea, which are usually caused by humans.

Some unicellular algae, such as blue-green algae, seem to belong to plants because like green plants, they also have chloroplasts, which are able to carry out photosynthesis, synthesize carbon dioxide and water into their own nutrients, and then release oxygen. However, careful study has found that these algae do not have the cell nuclei of green plants, and are actually close relatives of bacteria. They are also prokaryotes.

Different from prokaryotes, eukaryotes have real cell nuclei. The genetic materials in eukaryotes are wrapped by nuclear membrane. There are many pores in the nuclear membrane. Through these pores, the materials in the nucleus can communicate with the materials in the cytoplasm outside the nucleus.

The simplest organisms with cell nuclei are unicellular organisms, which can be divided into unicellular animals and unicellular plants. Unicellular plants, such as unicellular green algae, are able to carry out photosynthesis and live on their own, just like ordinary green plants. Unicellular animals are also known as protozoans, among which amoebas, paramecia, and ciliates are more common. Amoebas live in ponds, paddy fields, or ditches, with small bodies that are colorless and transparent. The largest amoeba is 0.2–0.4 millimeter in diameter and can be seen by the naked eye. But, to further observe, you have to use a microscope. There is only a thin membrane on the surface of the body of an amoeba. Inside the membrane is a relatively transparent and uniform cytoplasm, also known as protoplasm. The cytoplasm can flow continuously, and the shape of the cell membrane changes with the flow. This may be why this animal is called amoeba.

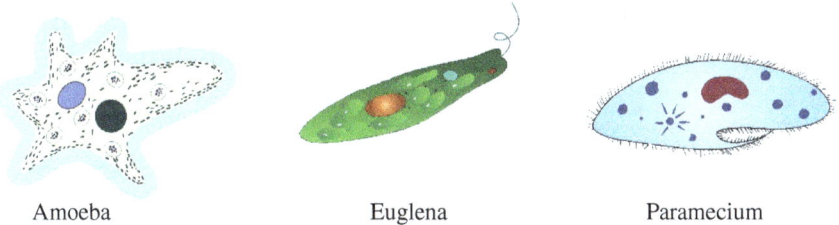

| Amoeba | Euglena | Paramecium |

Different kinds of unicellular organisms

When the amoeba carries out the movements of deformation, some finger-like protrusions of different lengths will appear on the surface of the cell, and the whole body moves along the direction of the protrusions, so these finger-like protrusions are called pseudopodia. Besides movements, another function of pseudopodia is to catch food. It can reach out to the food, wrap the food into its body, form food bubbles, and then digest the food. Amoeba has no gender difference, neither male nor female. It reproduces by dividing its original body from one into two, which is similar to bacteria.

Although amoeba is a single cell, it has the life characteristics of all animals that can live independently, such as response to stimulation, movement, predation, growth, and reproduction, so it is a kind of lower animal.

There is only one cell in the unicellular organism, four cells in a Gonium, and 16 cells in a pandorina. Generally, the higher class the organism is, the more cells there are. It is estimated that there are 2 trillion cells in a newborn baby.

Higher organisms are multicellular. In the process of long-term evolution, different cells have functional divisions of labor, and the shapes of the cells are more diversified. The sperm cells of animals are like tadpoles with long tails, which is convenient for sperm cells to swim in the reproductive tract and enter the egg cells for fertilization. The red blood cells are disc-shaped, which greatly increases the surface area and facilitates the gas exchange of carbon dioxide and oxygen. Nerve cells have long axons, some even more than 1 meter, just like telephone lines, which is also adapted to its function of transmitting nerve impulses. The shapes of higher plant cells vary greatly with their functions. The supporting and conducting cells at the base of plants are often strip-shaped, while the guard cells in the leaf epidermis are in the shapes of half-moons. Every two cells form a stoma to facilitate respiration and transpiration.

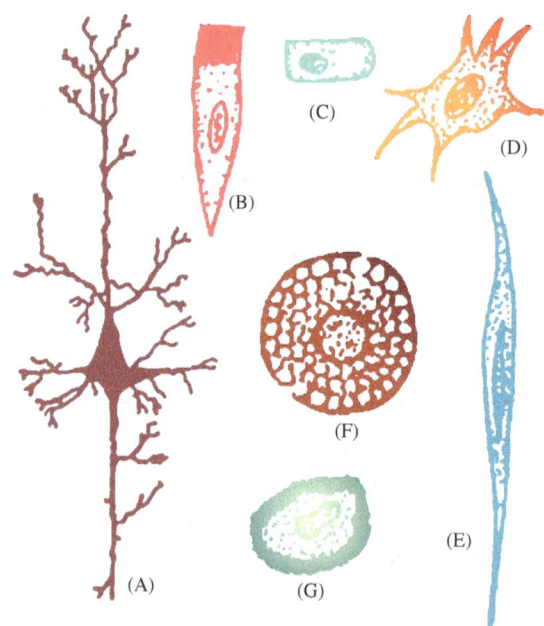

The morphology of the cells

(A, C, D) Epithelial cells; (B) connective tissue cell; (E) muscle cell; (F) egg cell; (G) nerve cell.

On the whole, the cells are small and nearly spherical, so as to ensure a relatively large surface area, which is conducive to metabolism and resistance to severe environmental conditions such as drought and freezing. However, there are exceptions. The unfertilized egg of birds contains only one cell, but it is very large.

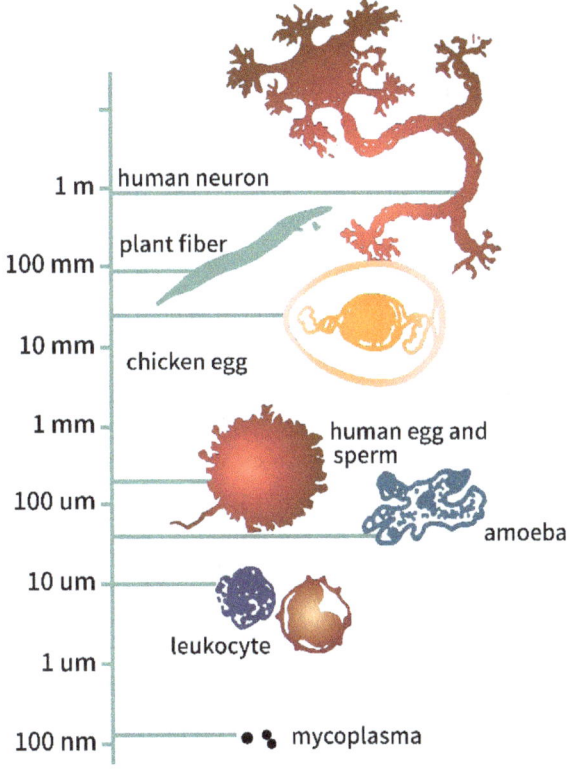

human neuron

plant fiber

chicken egg

human egg and sperm

amoeba

leukocyte

mycoplasma

The sizes of the cells

The egg cells of ostrich can reach 7–8 centimeters in diameter, which are the largest animal cells in the world. Why? It turns out that there are a lot of yolks in the eggs of birds, which are the nutrients for embryonic development. Only when the eggs are large enough can they provide sufficient nutrition for the development of embryos.

Cells can be divided into prokaryotic cells and eukaryotic cells. Both plant cells and animal cells are eukaryotic cells. The main difference is that plant cells have cell walls.

3. Cell Wall, the Protection Coat

The outermost layer of plant cells is a thick hard wall called cell wall. It is one of the important characteristics that makes plant cells different from animal cells. So, how does this seemingly luxury cell wall form? What is the special significance for plant cells?

Scientists have found that the cell wall of plants is produced by the secretion of cells, which can be divided into three layers: the new, thin wall called primary wall; the later thick wall with stripes, called secondary wall; and the materials between two cells, called middle layer, which makes the cell wall adhere and reduces the pressure between cells. The main components of cell wall are cellulose. Besides, there are also the hemicellulose, pectin, and lignin.

Lignin only exists in the mature cell wall, which makes the cell wall hard and protects the cell from external mechanical damage. When the cell is alive, the cell wall can deteriorate due to the penetration and accumulation of other substances. For example, the cell walls of rice and wheat contain silicates, which can resist lodging. Therefore, the cell wall can maintain the shape of the cell and protect the cell.

Some microbes also have cell walls, such as fungi, yeasts, and bacteria. The cell wall of bacteria contains no cellulose. For bacteria, cell wall is not a luxury, but a necessity. According to the staining reaction to a purple dye, bacteria can be divided into Gram-positive bacteria and gram-negative bacteria, the former such as Staphylococcus and Neisseria gonorrhoeae, and the latter such as *Escherichia coli* and typhoid bacillus. Gram-positive bacteria can be stained with this dye, while Gram-negative bacteria are not stained or slightly stained. This is due to different cell wall structures.

The cell wall of Gram-positive bacteria is thick, which consists of 15–50 layers of peptidoglycans. In addition, there are also proteins and polysaccharides. However, the cell wall structure of Gram-negative bacteria is more complex, which can be divided into two layers. The inner layer is a peptidoglycan layer, which is thin and connected with the outer layer through lipoprotein. The outer layer is called outer membrane, which is basically a layer of phospholipid and protein membrane. The outer membrane contains lipopolysaccharides and lipoproteins, which are related to the toxin activity of bacteria, and are also the cause of fever after the bacteria invade the human body.

Penicillin is a familiar antibiotic. Its action mechanism is to kill bacteria by preventing the synthesis of peptidoglycans in the cell wall. Therefore, it has better killing effect on Gram-positive bacteria containing a large number of peptidoglycans in cell wall, but it is relatively poor against Gram-negative bacteria.

Some bacteria have a layer of fibrous material outside the cell wall called capsule. A capsule is secreted outside the cell wall by bacteria, which is composed of polysaccharides and proteins, containing neutral saccharides and phosphoric acids, but without sulfuric acids. The capsule is not strictly necessary for the survival of bacteria, but it can protect bacteria against adverse environment and greatly enhance the toxicity.

The main component of the cell wall of fungi and yeasts is chitin, also known as chitosan, a kind of polysaccharide.

Although animal cells have no cell wall, some cells have dendritic polysaccharide chains formed by glucose, galactose, arabinose, and so on. Like antennas, these polysaccharide chains play important roles in the recognition and communication among cells. Furthermore, the human ABO blood type is also related to the polysaccharide chains on the erythrocyte membrane.

4. Cell Membrane, the Exchange Channels

From animals, plants to microorganisms, all cells are covered with a kind of membrane. This membrane is known as cell membrane that can separate the cell from the surrounding environment. The basic components of cell membrane are proteins, lipids, and saccharides. In the cell membrane, lipid molecules are arranged in two layers, and proteins are embedded in, across, over, or attached to the lipid bilayers. As for saccharides, they are combined with proteins or lipids.

Under the electron microscope, the cell membrane is the three-layer structure like a sandwich, with two dark layers and one bright layer. The membrane with this kind of structure is also called unit membrane. The inner and outer dark layers are proteins, and the brighter layer in the middle is lipids.

Cell membrane plays an important role in controlling the material exchange between the cell and the external environment. In the cell membrane, the hydrophobic end of lipid molecules faces inward, while the hydrophilic end faces outward, so that the bilayer lipid molecules are firmly bound together by hydrophobic force. The small molecular substances and fat-soluble substances, such as glycerin, water, oxygen, nitrogen, benzene, and urea, and can diffuse from the high concentration side to the low concentration side through the cell membrane. However, some hydrophilic substances, such as glucose, amino acids, nucleotides, and all ions, cannot pass through the lipid bilayer freely. However, there are special transport proteins on the cell membrane. When these substances are combined with transport proteins, the conformations of transport proteins are changed, thus transporting these substances to the other side. There is also a protein called $Na^+ - K^+$ ATPase on the cell membrane. When it is stimulated, its conformation is changed. It works like a pump, Na^+ ions are transported out of the membrane, while K^+ ions are transported into the membrane to maintain a certain potential of the cell membrane. This potential is necessary for cell transport of certain substances and nerve conduction.

Some larger substances can enter and leave the cell membrane through phagocytosis or exocytosis. The process of endocytosis of larger solid particles such as bacteria is called phagocytosis, and when the phagocytic material is in the form of solution or very small particles, it is called pinocytosis. Phagocytosis is common in unicellular animals, but is also found in multicellular animals. For example, leukocytes and macrophages in human blood have strong phagocytic capacity, which can phagocytize the bacteria that invade into the body. When the human body is injured, leukocytes and macrophages will accumulate around the wound and begin to phagocytize bacteria. When these cells swallow a large number of bacteria, they will die due to overeating, and the wound will fester. Pus is actually dead leukocytes. Of course, cells can also excrete substances that are not needed in the body, which is called exocytosis.

Besides material transport, another important function of cell membrane is to transmit information. There is a protein called "receptor" on the cell membrane. When it binds with the ligands in the external environment like bolts and nuts, it will trigger the structural changes of the proteins connected with the receptor, thus activating the specific enzymes on the cell membrane and transmitting the signal to the cell. For example, when the concentration of glucose in the blood is too low, the brain will secrete adrenaline, which follows the blood circulation to the liver cells, and then binds to the adrenaline receptor on the liver cells, thus activating a series of enzymes to break down the starch in the cells into glucose. In recent years, studies have shown that in the process of cell membrane signal transmission, if some links fail, the cell will become cancerous.

5. Vital Organelles in the Cytoplasm

Compared with bacteria, the structure of animal and plant cells is more complex. One prominent feature is the emergence of small organs in the cytoplasm through evolution. These small organs immersed in the cytoplasm are called organelles. They are some morphological entities surrounded by membranes that can perform specific life functions like the human heart, liver, and kidney.

Mitochondria are one of the most ubiquitous and important organelles in eukaryotic cells. They were first discovered in insect striated muscle cells by Swiss anatomist and physiologist Albert von Koelliker in 1857. Other scientists have found the same structure in other cells, confirming Albert von Koelliker's discovery. In 1888, Albert von Koelliker isolated the organelle. In 1897, the German scientist Carl Benda named this organelle mitochondria. As the name suggests, mitochondria are linear or granular in appearance.

Under an electron microscope, the typical mitochondrion looks like a sausage. It includes an outer membrane, inner membrane, outer chamber between the inner and outer membranes, and inner chamber wrapped by the inner membrane. The outer membrane has high permeability to various substances. Some people think that there are small holes in the outer membrane. The permeability of the inner membrane to substances is very low, and it can only let some non-charged small molecules pass through, such as water and pyruvic acid. The inner membrane folds inward to form wrinkles or tubules, called crista. The existence of crista greatly enlarged

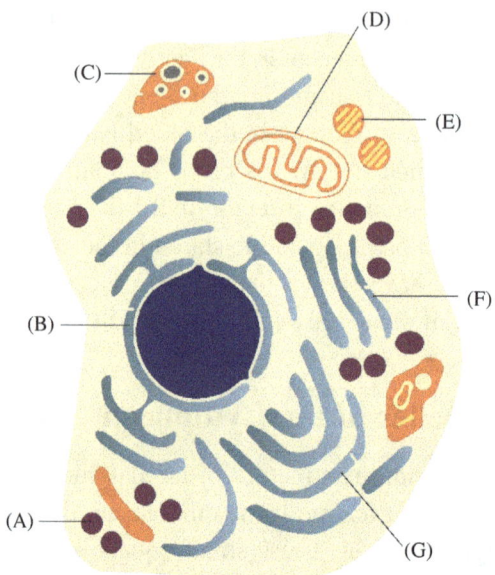

Organelles in animal cell

(A) Vesicle; (B) nucleus; (C) lysosome; (D) mitochondria; (E) peroxisome; (F) Golgi; (G) endoplasmic reticulum.

the surface area of inner membrane and improved the metabolic efficiency. On the crista, there are many small stalked particles called ATP (adenosine triphosphate) enzyme complex, which is the site of ATP synthesis in mitochondria. ATP is an unstable high-energy compound, which can release more energy during hydrolysis, and is the most direct energy source for organisms. The inner chamber is a liquid matrix containing ribosomes, DNA, RNA, and enzymes. Ribosomes can synthesize various proteins; DNA and RNA are nucleic acids, which can carry much genetic information; enzymes can catalyze a wide variety of biochemical reactions in the organism at all times, and these biochemical reactions are called metabolism, including material metabolism and energy metabolism. Metabolism is the most basic characteristic of life. Without metabolism, there will be no life. Therefore, enzymes are very important as biocatalysts.

Mitochondria are involved in the oxidation of intracellular substances and respiration. The decomposition products of saccharides, protein, and fat in cells are completely oxidized in the mitochondrial matrix, and then the released energy is stored in ATP through the respiratory chain on the inner membrane of mitochondria. Most of the energy consumed by intracellular substance transport, muscle contraction, and nerve conduction is provided by the ATP synthesized in mitochondria. For this reason, mitochondria are known as the power plants of cells, just like power plants that provide energies for human daily lives.

Interestingly, mitochondria are similar to bacteria in size, shape, ring chromosome, and division mode, so some people think that it may be evolved from the bacteria living in the cells. When fertilized, the egg receives only the nucleus of the sperm, so in the fertilized egg, the cytoplasm is completely from the mother side, and the mitochondria are maternally inherited. This is used in forensic medicine for paternity testing.

Chloroplasts are unique organelles of plants and are the places where plants carry out photosynthesis and capture solar energies. In the leaves of green plants, there are different numbers of chloroplasts. As far as the higher plants are concerned, the shape of chloroplast is like a convex lens, of which the outermost is a two-layer smooth unit membrane, with a gap between the inner and outer membranes. The inner membrane is filled with a kind of liquid matrix, in which there are many disc-shaped thylakoids that are stacked together and much like a pile of coins. These stacked thylakoids form grana. The thylakoids constituting grana are called grana thylakoids. The large thylakoids that cross over the chloroplast matrix and across through two or more grana are called stromal thylakoids. Therefore, thylakoids have two kinds. There are DNA, ribosomes, and many enzymes in thylakoids.

The main function of chloroplasts is photosynthesis, which is to use solar energy to synthesize carbon dioxide and water into carbohydrate and release oxygen. Photosynthesis can be divided into two stages: light reaction and dark reaction. Light reaction is the process of chlorophyll and other pigment molecules absorbing and transferring light energy, converting light energy into chemical energy, and forming ATP and reduced coenzyme II (NADPH). In this process, water molecules are decomposed and oxygen molecules are released. Dark reaction is the process of making glucoses and other nutrients by using the intermediate products formed by light reaction. As a result of photosynthesis, light energy is converted into chemical energy and stored in carbohydrates for human and animal use.

Lysosome was discovered in 1955. It is the "digestive organ" of cells, and contains more than 50 kinds of hydrolases, including lipases, proteolytic enzymes, and nucleases. These enzymes are all acid hydrolases, and the optimum pH value is 5.0. There are two kinds of lysosomes: primary lysosome and secondary lysosome. Primary lysosome is the newly formed lysosome in the cell, and has a vesicular structure, in which the enzymes are in a latent state. Secondary lysosome is the digestive vesicle formed by the combination of primary lysosome and digestive products.

Lysosomes have important functions in cells. First, they can digest the macromolecular nutrients, bacteria, and viruses that have been phagocytized into cells. The digested nutrients are used by cells, and the remaining residues are discharged from the cells, thus playing roles of nutrition and defense. Second, when some organelles are aging, they can be surrounded and cleared by lysosomes, which is conducive to the renewal of these organelles. When animals are hungry, lysosomes can surround and digest some of their own non-essential substances in the cells of their bodies for nutrition, so as to renew the necessary components and avoid permanent death of animals. Third, in the process of animal growth and development, the excess organs are removed. For example, in the late development of tadpole, the lysosomes in tail cells will break down by themselves, and the released hydrolases will dissolve the tail cells, thus making the whole tail disappear.

Although there are many kinds of enzymes in lysosomes, the types of enzymes in each lysosome are limited. Once the membrane of primary lysosome breaks down, the released hydrolases will play a powerful role in digestion, which can digest the whole cell and even affect the surrounding tissues.

Lysosomes exist in animal cells, plant cells, and protozoan cells. However, there are no lysosomes that exist alone in plant cells, only some small bodies with different substances. These small bodies have different names because of their different substances, such as spherosomes, aleurone grains, and proteoplasts. These small bodies contain acid hydrolases, for example, there are lipases, acid phosphatases, and other hydrolases in spherosomes. In addition, the vacuoles of plant cells also contain acid hydrolases. There are no single lysosomes in bacterial cells, but there are gaps between the cell walls and the cell membranes, which contain hydrolases, which act as lysosomes.

Of course, there are other important organelles in animal and plant cells, such as endoplasmic reticulums, Golgi apparatuses, centrosomes, and peroxisomes. Endoplasmic reticulums are tubular or flat cystic organelles surrounded by membranes in cytoplasm. It has two kinds of endoplasmic reticulums: rough endoplasmic reticulum and smooth endoplasmic reticulum. Its functions involve protein and fat synthesis, material transportation, and detoxification. Golgi apparatuses are another kind of organelles surrounded by flat sacs, branching tubules, and vesicles. It is related to protein processing, the formation of hydrolases in lysosomes and the formation of cell walls in plants.

In 1953, when E. Robinson and R. Brown of Britain observed plant cells with an electron microscope, they found a kind of granular substance. In 1955, Palade also observed similar particles in animal cells. In 1958, Roberts named this particle ribonucleoproteasome or ribosome for short. It is a type of ubiquitous particle in cells. The ribosomes in eukaryotic cells are a little larger than those in prokaryotic cells, but the ribosomes in mitochondria and chloroplasts are as large as those in prokaryotic cells.

The ribosome consists of two subunits that look like potatoes. There is a pit on the surface of the nearly spherical large subunit, while the small subunit is slender and has a circle of grooves. Ribosomes are machines that synthesize proteins in cells. It is well known that proteins are the executor of life activities in cells, such as various enzymes that catalyze metabolisms. Although it is now found that only a small number of enzymes in nature are ribonucleic acid, various hormones that regulate physiological functions, and hemoglobins that transport oxygens and carbon dioxides in the blood, are all proteins. It can be said that there will be no life without proteins.

Ribosomes exist in cytoplasm; some are attached to nuclear membrane or rough endoplasmic reticulum, and others are dissociative. When a cell

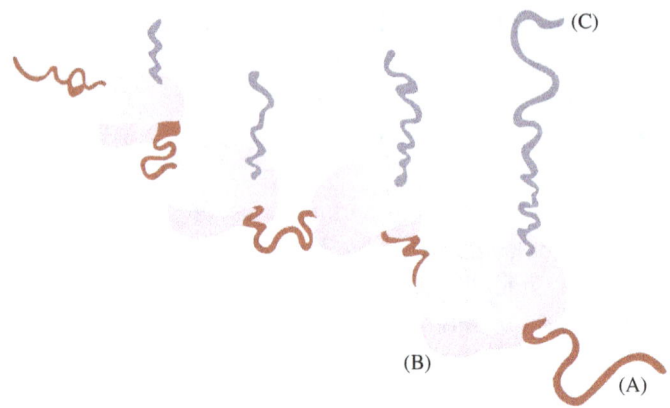

Ribosomes

(A) Messenger RNA; (B) ribosome; (C) protein.

needs to synthesize a specific protein, it will first transcribe a messenger RNA from the DNA that stores the specific life information, and then the messenger RNA will combine with the ribosome, and then synthesize the protein. In cells with vigorous protein synthesis, multiple ribosomes can be seen on one messenger RNA at the same time. In this way, a large number of proteins can be synthesized in a short time to meet the needs of life activities.

6. Cell Nucleus, the Nerve Center

Another major feature of animal and plant cells that are different from bacterial cells is cell nucleus, and it is an important organelle for storing genetic information in cells, although there is very little genetic information in mitochondria and chloroplasts of plant cells, mitochondria of animal cells, and bacterial plasmids (circular genetic materials in bacterial cells).

As early as the 17th century, Antonie van Leeuwenhoek, a Dutch optician, discovered the cell nucleus with his self-made microscope. In 1831, Scottish botanist R. Brown first used the word "cell nucleus" and believed that all cells have cell nuclei. Modern biological studies have shown that besides bacteria, actinomycetes, and cyanobacteria, other types of living cells have cell nuclei at a certain stage or in the whole life cycle.

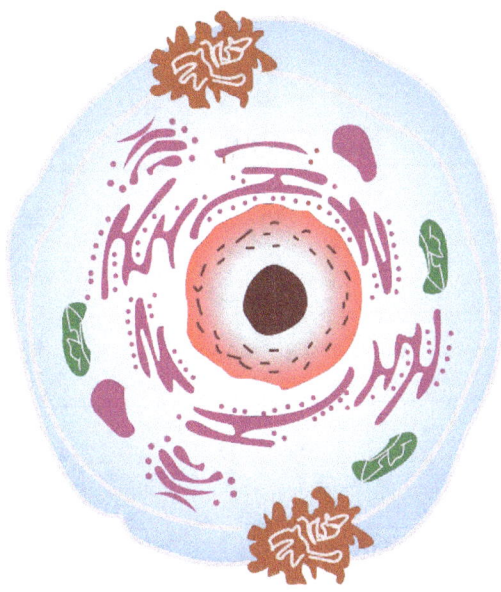

Cell nucleus (central part of the cell)

Generally speaking, eukaryotic cells will die soon after losing their cell nuclei. Only a few cells can survive without cell nuclei. For example, mature red blood cells in mammals can live for more than 120 days after losing their cell nuclei; nutrient transport cells in the phloem of plants can perform functions for many years in the absence of cell nuclei. But, in general, cells cannot live without cell nuclei.

A cell nucleus is usually located in the center of the cell and consists of a double nuclear membrane. There are some holes in the nuclear membrane through which the cell nucleus and cytoplasm can exchange substances with each other.

The main substance in cell nucleus is the chromosome, which is composed of DNA, protein, and a small amount of RNA. DNA double helix is highly compressed in the chromosome (compression ratio ≤10,000). Cell structure and the main information controlling cell growth, development, physiological activities, and reproduction are stored in DNA, so the cell nucleus is the "supreme headquarters" or "nerve center" of the whole cell.

Chapter 2

The Bright Prospects of Cell Culture

1. Tissue Culture for Micropropagation Seedlings

From luxuriant little grasses to towering trees, and from fruits and vegetables to all kinds of crops, they are all developed from single fertilized egg cells. So, can we just take a cell from the plant body and culture it into a complete plant? A century ago, it was a beautiful fantasy, but now it can be done through plant tissue culture.

What is plant tissue culture? In short, it is the process of culturing plant tissues and organs or further developing into complete plants by controlling nutrition, light, temperature, humidity, and other environmental conditions under strict sterilization circumstances.

The history of plant tissue culture can be traced back to 1902, when the famous German botanist Gottlieb Haberlandt predicted that plant cells have totipotency. The so-called "totipotency" means that every cell in plants has a variety of potential abilities, which can accomplish the whole process from cell proliferation and differentiation to complete plant development. In plants, the talents of these cells are buried, and they can only undertake a specific task silently, such as forming stem tissue, root tissue, and leaf tissue. However, due to technical limitations, the cells cultivated by Gottlieb Haberlandt did not divide.

In 1904, Hanning successfully cultivated the normal development embryos of radish and cochlearia in the medium, and the embryos are the key components of seeds; he became the originator of plant tissue culture technology. After the 1930s, the plant tissue culture technology made great progress. The founders of plant physiology in China, Jitong Li, Zongluo Luo, and Shiwei Luo, successively found in experiments that the extracts from Ginkgo biloba endosperms and young mulberry leaves can promote the growth of isolated Ginkgo biloba embryos and maize roots, respectively. Therefore, it is believed that vitamins and some organic substances are indispensable components in plant tissue culture.

In 1934, White, an American, successfully established the first plant tissue that could grow infinitely with tomato roots. In 1956, Miller found that kinetin could strongly induce the callus to differentiate into buds. This is important progress in plant tissue culture. Two years later, Steward and others successfully cultured complete plants with carrot cells, which confirmed the totipotency of plant cells and opened up a new technical field. Since then, plant tissue culture technology has developed rapidly in the world. Up to now, nearly 1000 species of plants have been able to propagate rapidly by such means.

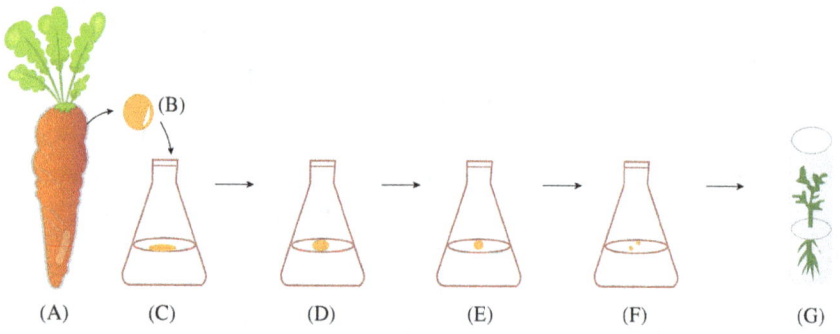

The process of carrot regeneration through explant culture

(A) Carrot; (B) carrot cross section; (C) carrot cell culture; (D) cell dedifferentiation to form callus; (E) callus cells divide and differentiate into embryoids; (F) embryo; (G) plant seedling.

How is plant tissue culture carried out? The following is a brief description of several steps.

First, select suitable explants. What are explants? In plant tissue culture, in order to achieve the purpose of rapid propagation, plant organs or tissue slices are often selected as the culture object, such as a small part of a bud, stem, and leaf, which are called explants. Choosing a good explant is the beginning of successful culture. When the explant is an old tissue or organ, its ability to develop into the whole plant will be weakened, so one should pay attention to select a young tissue or organ, which can easily produce a large number of calluses. The so-called callus originally refers to the newly grown tissue around the wound after the plant is injured, and in plant tissue culture, the callus refers to the cell mass, which is grown out from the culture material and has the ability of producing offspring. A complete plant can be regenerated from the callus. In addition, when choosing explants, attention must be paid to the health of their appearances. The explant should not be too small, and should have more than 20,000 cells (that is, 5–10 milligrams), so it can easily survive.

Second, sterilize the explants. Explants often contain bacteria and other microorganisms. If they are not disinfected, the bacteria will multiply during the culture. Because bacteria propagate much faster than cells, they will exhaust the nutrition of cells in the culture medium, so that the culture is full of bacteria, and the explants grow slowly or die due to lack of nutrition. Moreover, the pathogens can propagate using cells as nutrition, which directly leads to explant decay. It is necessary to disinfect the explants. When disinfecting, attention should be paid to choose the appropriate disinfectant. The disinfection time of young materials is shorter than that of mature materials.

Third, prepare culture medium. In the natural state, plants grow in the soil. The tissue culture is carried out indoors, and the artificial medium replaces the natural soil, and the culture medium is more nutritious than the soil. Although there are many kinds of media for plant tissue culture, they usually include three major kinds of components: (1) abundant basic components, such as nitrogen, phosphorus, potassium, sucrose, or glucose (generally up to 30 grams per liter); (2) trace inorganic substances, such as iron, manganese, and boron; and (3) trace organic substances, such as kinetin, indole acetic acid, and inositol. Due to the different culture purposes, the contents of kinetin and indole acetic acid in various media varied greatly. The function of indole acetic acid is to promote cell growth, and the function of kinetin is to promote cell division. When the content of indole acetic acid is higher than that of kinetin, it is beneficial to induce the production of calluses from explants.

Fourth, induce explants to grow calluses. Explants are pieces of tissues or organs that tend to mature and stereotype. If they are cultivated into whole plants, they must be rejuvenated. Adding higher concentration of auxin to the culture medium can make the cells in the explants get rid of the lazy state and start vigorous growth again so as to develop calluses. The explants can be cultured in solid culture medium. The sterilized explants can be inserted or pasted into the solid culture medium. The advantages of solid culture are simple, multi-layer culture and small area. However, there are also disadvantages, that is, the nutrient absorption of explants in solid medium is uneven, and the harmful substances produced in the process of cell growth are not easy to diffuse out. These defects can be avoided if the explants are placed in a liquid culture medium, but an oscillator is needed. Through culture, the explants grow out calluses. Since the cells are growing with ready-made nutrients, light is not needed in principle.

Fifth, improve the nutrition of the callus. After 4–6 weeks of rapid cell division, calluses finally grow in the medium. But, at this time, the water and nutrients in the medium have been exhausted, and a large number of toxic substances have been accumulated. It is necessary to transplant them in time to improve the nutritional environment. After transplantation, the cells in the callus begin to expand vigorously, which is conducive to rooting and germination.

Sixth, a callus grows out roots and buds. The callus is transplanted into fresh culture medium containing appropriate amount of cytokinin and auxin to induce embryoid formation. The so-called embryoid is a structure formed in tissue culture with bud end and root end, similar to an embryo in seeds. If the embryoid is further developed into a complete plant, light is needed.

Seventh, transplant the plantlets. The plant seedlings cultivated in glass bottles should be transplanted outdoors in time to facilitate their growth.

Plant tissue culture is a very practical technique. According to calculations, a 20–square-meter greenhouse can accommodate hundreds of thousands of test tube seedlings. In addition, tube propagation can be carried out throughout the year without seasonal restrictions. This technology is widely used.

One of the main applications of plant tissue culture technology is factory-like rapid seedling, in order to propagate those rare flowers and fine crop varieties which are difficult to propagate by conventional methods, especially the expanded culture of excellent mutant plants.

Relying on this technology, many distinctive flower industries have been formed and huge economic benefits have been created. Since the 1960s, the orchid industry based on the rapid propagation technology by tissue culture has benefited some countries in Europe, America, and Asia. Singapore and Thailand can make tens of millions of dollars a year just by exporting orchids. In addition to orchids, lilies, chrysanthemums, anthuriums, Boston ferns, and other flowers have reached an annual output of more than 1 million plants. Other promising breeding targets include carnation, narcissus, gladiolus, tulip, Clivia, Actinidia chinensis, seedless watermelon, and hawthorn.

In the early 1970s, China began to research and apply tissue culture technology, and is currently at the leading level in the world in this field. In the past 10 years of the 21st century, high-tech tissue culture technology has become common practical technology and has been widely used. In some agricultural and forestry universities and large biotechnology companies, there are tissue culture rooms. In some areas, the industrialization of flower tissue culture seedlings has become a large-scale production. The annual output of tissue culture seedlings of foliage plants produced by the Guangzhou Flower Research Center is more than 10 million plants. The Flower Research Center of Horticulture Research Institute of Yunnan Academy of Agricultural Sciences has built a tissue culture room with an annual production capacity of 50 million plants. The Yunnan Yuxi high-tech Development Zone has realized the large-scale production of tissue culture seedlings of tropical orchids, and the Biotechnology Center of Hunan Provincial Forest Botanical Garden has realized the specialized, large-scale, and commercialized production of eucalyptus test tube seedlings, with an annual production capacity of millions. It is the first time to explore the industrialization development road of eucalyptus test tube seedlings in China. According to the preliminary statistics, there are more than 100 families and more than 1000 species of plants that can be rapidly propagated with the help of tissue culture technology. However, the plants that have really accomplished large-scale industrialized tissue culture production mainly include the crops, flowers, fruit trees, vegetables, and traditional Chinese medicines with important economic values. More than 10 rapid propagation production lines for grape, apple, banana, potato, sugarcane, orchid, and eucalyptus have been built in China, and more than 100 million tube seedlings are supplied annually, among which the banana tube seedlings have entered the international market.

Another major application of plant tissue culture technology is virus-free plants. At present, there are more than 500 kinds of plant virus diseases, among which the most seriously damaged are food crops (rice, potato, sweet potato, etc.), economic crops (rape, lily, garlic, etc.), and flowers (carnation, orchid, iris, etc.). These viruses cause serious yield reduction and variety deterioration of plants. For the virus, so far, there is no effective treatment in the world. Using plant tissue culture technology, virus-free plantlets can be established quickly. Generally speaking, the stem tip area of a plant is virus free. The stem tip material for virus-free culture is very small, generally 0.1–0.3 mm. Not only does the operation need the help of a microscope and other instruments but there are also many difficulties in culturing such small stem tip materials, and the survival rate is quite low. But, it is very effective in preventing viruses.

The plants that have been put into production of non-toxic seedlings include potatoes, orchids, chrysanthemums, lilies, strawberries, and garlics, and many countries have established production bases of non-toxic seedlings. The rapid propagation of virus-free seedlings of strawberry in Japan can increase the yield by 30–50%. Some European fruit-producing countries also widely use the virus-free seedlings, which greatly increase the yield and improve the fruit quality.

In China's Guangdong, Guangxi, Hainan, Yunnan, and other bananas-producing provinces, a big problem is often encountered, that is, once the banana is infected by viruses, not only will the yield be reduced but also the varieties will be seriously degraded. Within a few years, the banana tree will lose its production capacity. By using plant tissue culture technology, we can select excellent banana varieties, establish non-toxic cell lines, and propagate seedlings *in vitro*. After the seedlings grow to a certain extent, they are moved to the field for further growth. As the seedlings are completely sterile in the process of rapid propagation, they are also non-toxic after being transferred into the field, which can greatly reduce the disease rate and increase the yield. After a few years, the rapid propagation seedlings can be replaced once so as to ensure that the varieties do not degenerate for a long time. At present, several provinces in southern China are using this method to produce bananas, and the results are very good.

In terms of rapid propagation and virus-free plants, the Heilongjiang Provincial Seed Company and other units have built a production base of non-toxic potato seedlings. The production of non-toxic seedlings through tissue culture can increase the yield of potatoes by more than 50%, and successfully prevent potato degradation. More than 40% of strawberry seedlings can be obtained by Shanghai Academy of Agricultural Sciences. Liuzhou, China, has also built a production base of sugarcane non-toxic seedlings.

Moreover, plant tissue culture technology is also of great significance in saving endangered species. For some endangered rare plant species, even if there is only one plant left, it can be allowed to reproduce a large number of offspring in a short period of time through tissue culture technology, so as to alleviate the endangered situation and enrich the natural treasure house of species.

2. Plantlet Regeneration with Protoplasts

A small piece of plant tissue or organ cultured in a test tube or glass bottle can rapidly regenerate complete plants, so can the culture of any plant somatic cell regenerate a complete plant? In theory, it can. Because of the totipotency of plant cells, except for a few somatic cells without cell nuclei such as nutrient transport cells in plant phloem, each somatic cell contains all the genes required for the life activities of the plant to which it belongs. As long as the conditions are suitable, all somatic cells can be used as seed cells and have the potential to develop into complete plants.

But, unlike animal cells, plant cells are surrounded by a hard and elastic cell wall. This thick cell wall makes plant cells very lazy and hinders the full play of totipotency. If this layer of cell wall is removed, perhaps the plant somatic cells will be like seeds that will grow small seedlings if planted. The question is how to remove the cell wall.

As early as 1892, biologist Cole Koehler first removed the cell walls of algae through the mechanical method. He had placed the cells in a high concentration of sugar solution, and as a result, the cell walls and the cytoplasms were separated. The cell walls were finally broken, and the bare cells were obtained. These naked plant cells are also called protoplasts. Later, many scientists tried to prepare protoplasts with this method. Unfortunately, the yield by using this method was extremely low, and the protoplasts of many cell types could not be obtained in this way.

In the early 1960s, Professor Cocking at the University of Nottingham invented a method for large-scale preparation of protoplasts of higher plants, which is ingenious and novel. Cocking digested the cell walls of root tip cells of tomato seedlings with cellulase and obtained a large number of protoplasts. The advantages of the enzymatic method are that the protoplasts can be produced in a large quantity and are not easily broken. At present, with this method, a large number of protoplasts can be isolated from any part of plants, such as leaves, flowers, fruits, roots, tumor tissues, and calluses or cells cultured *in vitro*. Generally speaking, the genetic traits of protoplasts prepared from mesophyll tissues are relatively consistent, but the tissues or cells cultured *in vitro* have great differences in both genetic traits and physiological status, so it is better not to use them for seedling raising.

Due to the advantages of enzymatic protoplast preparation, this method had been greatly developed at that time and is still widely used in the field of plant cell engineering in the 21st century. In practice, cellulase, pectinase, snailase, and callase are commonly used. Cellulase is a compound enzyme preparation extracted from a fungus called *Trichoderma viride*. Pectinase is extracted from another fungus, Rhizopus, which can separate cells from tissues. Snailase and cellulase have better digestion effects on pollen mother cells. Some enzyme preparations contain many impurities, such as phenols, nuclease, protease, and peroxidase. These impurities not only reduce the enzyme activities, but also have toxic effects on the protoplasts. In practical application, it is better to select high-purity enzyme preparation.

After a large number of viable protoplasts are obtained, they can be cultivated in culture medium. The protoplasts can be cultured by the solid culture method, which is that the prepared protoplasts are evenly fixed in agar culture medium. In the specific operation, the solid culture medium with double concentration was first prepared, then heated and dissolved, cooled to 45°C, mixed with the same amount of protoplast suspension that had already prepared, poured into a glass culture dish with a diameter of 6 centimeters, then shaken quickly and gently to make the protoplasts evenly dispersed in the agar culture medium, sealed with gummed tape, and then put into a glass culture dish with a diameter of 9 centimeters, in which a wet sterile filter paper is placed to maintain a certain humidity. The advantage of this culture method is that it is convenient to observe the process of growth and development of a protoplast. This method was used for the first successful cultivation of tobacco mesophyll protoplasts.

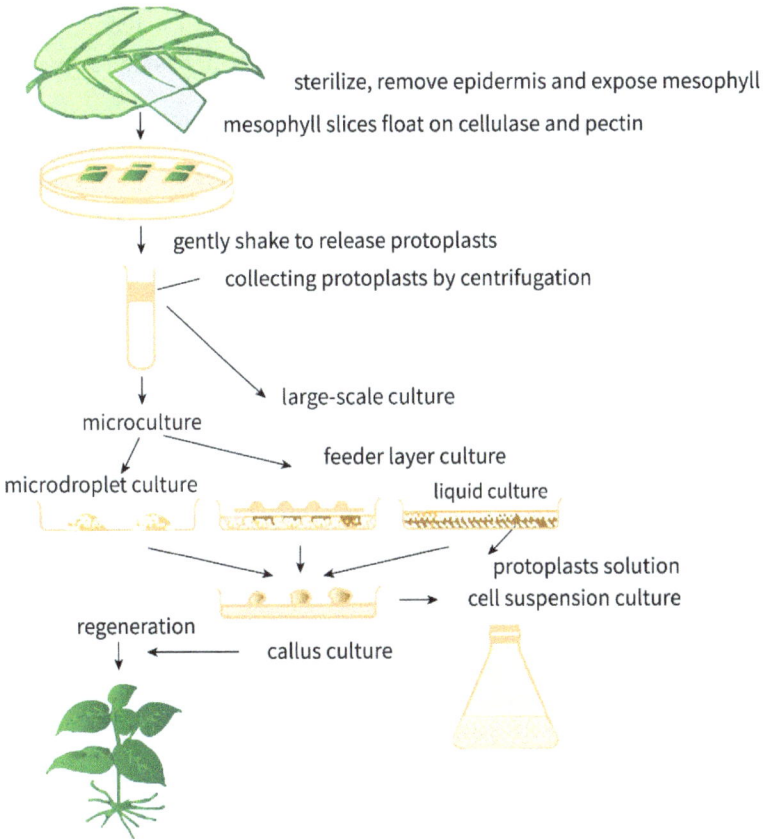

Regeneration of plant seeding by protoplast culture

Protoplasts can also be cultured by the liquid culture method. This involves suspending the purified protoplasts in a liquid culture medium, which could be shallow liquid culture or hanging drop culture. Because protoplasts are easy to precipitate to the bottom of culture flask, they need to be shaken several times a day to facilitate aeration. This method can make the protoplasts grow faster, but when the cells divide into cell clusters, they need to be transferred to solid culture medium so that the cells can continue to proliferate or can be induced to differentiate into seedlings.

Protoplasts can also be cultured by the double-layer culture method. The protoplasts are first suspended in liquid culture medium and then transferred to solid culture medium. The combination of liquid and solid culture medium can keep good humidity, and fresh culture medium should be added periodically during the culture process, which is more conducive to the growth of protoplasts.

In the process of protoplast development, cell wall regeneration is the first step. The rate of cell wall regeneration is different for the different plants that the protoplasts come from; for example, in some plants of Vicia, the cell walls begin to synthesize 10–20 minutes after the protoplasts are isolated, while the protoplasts from tobacco leaves do not start to synthesize cell walls until 3–24 hours. In addition, the synthesis rate of young plant protoplasts is faster than that of mature plant protoplasts. After growing out cell walls, protoplasts become new plant cells and form cell masses after multiple divisions. The cell masses continue to divide and proliferate to form calluses. After induction, the calluses grow buds and roots, and then develop into plants.

So far, many plant protoplasts such as carrots, rape, potato, rice, wheat, cassava, strawberry, and apple have been successfully regenerated, and much important progress has also been made for the regeneration of fungi protoplasts and bacterial protoplasts. In 2013, Yu Wang and others reported the protoplast regeneration of Ganoderma lucidum. In 2015, Yuexia Lu and others reported the protoplast regeneration of *Coprinus comatus*, Ling Sun and others reported the protoplast regeneration of denitrifying phosphorus accumulating bacterium N14, and Zixuan Yang and others reported the protoplast preparation and regeneration of *Lactobacillus rhamnosus*. In 2016, Ruibo Jia and others reported the preparation and regeneration of protoplasts of Monascus sorghum M-3. In 2019, Guanghuan Li and others reported the preparation and regeneration of protoplasts of *Pleurotus abalonus*. These research results have potential application values.

3. Pollen Culture to Obtain Fine Breeds

In plant breeding, sometimes we would encounter a very difficult problem. After the seeds obtained by sexual hybridization are planted, the heredity of the hybrid plants that we get is unstable, and some good characteristics will be lost. What is going on here? We know that the traits of plants are controlled by genes, and genes exist in pairs in plant somatic cells. If A and a are the allelic genes controlling the height of the maize plant, and A is the dominant gene, when A exists, the maize will grow into a tall plant, and a is the recessive gene, that is, when a exists, the maize will grow into a dwarf plant. According to the law of free combination of genes, there are three genotypes AA, Aa, and aa in the offspring of the maize, and there are at least two phenotypes. Among them, AA and Aa are both tall plants and aa is a dwarf plant. However, Aa belongs to heterozygous genotype, and there will be separation of height traits in the offspring. Only after several generations of selective purification can a stable variety be formed. This has brought a lot of disadvantages to the breeding work, especially for the plants such as fruit trees, which have mainly asexual reproduction, highly heterozygous genotypes, and long growth periods. Depending on conventional breeding methods, it often takes many years to breed a stable new variety. However, this defect can be avoided by pollen breeding.

What is pollen? Pollen is the male germ cells of plants, in which the number of chromosomes is only one-half of normal somatic cells; thus, pollen is haploid, and a normal somatic cell is diploid. The normal pathway of pollen development is to produce the same haploid sperms, and then combine with the haploid eggs produced by the female reproductive organ through the pollination process to form diploid seeds. Pollen also has totipotency and can grow into a complete plant through culture. However, compared with normal diploid plants, haploid plants have many disadvantages such as small leaves, short plants, poor growth, weak vitality, and generally cannot blossom and bear fruit. In fact, there are haploid plants in nature, which are more common than haploid animals. In 1921, a haploid plant was first discovered in the higher plant Datura stramonium by A. D. Bergner.

Although haploid plants have little value in production, they can be used as an intermediate material in the breeding process. Moreover, because haploid plants have pure genes and no dominant genes can shield recessive genes, it is convenient for people to select recessive mutants with available traits, and the economic effect is very significant. In 1924, Blakeslee proposed the idea of using the haploid plants to get normal diploid plants by doubling the chromosomes in breeding.

Scientists have found a chemical that can double the number of chromosomes in a cell. It is colchicine, which is frequently used in breeding. If the pollen is artificially treated, that is, it is soaked in 0.2–0.4% colchicine solution, after 24–48 hours of treatment, and then cultured according to the conventional way, the chromosome number can be doubled, and the obtained normal diploid plants can blossom and bear fruit. The genotypes of the diploid plants are homozygous, and the characters of the improved varieties will not be lost in the future reproduction process. This can simplify the breeding process and shorten the breeding cycle. Moreover, the homozygous diploid after haploid doubling can show the recessive characters after homozygosity, which expands the selection range of traits, which is conducive to the design of crop variety improvement and mutation breeding.

In 1974, Nitsch and others pioneered the method of separating pollens by extrusion for culture. They removed the mature tobacco buds, placed them at 5° for 48 hours, disinfected the surface, and removed the anthers. The anthers were allowed to float in light on 28° liquid medium for 4 days of pretreatment. Then, the anthers were crushed with instrument to make pollen suspension. After filtration, centrifugation, and culture, only about 5% of pollen plants were obtained. The reason for the low success rate may be that the pollens lack the related substances needed for the initiation of cell divisions, such as the water-soluble anther factor for which the composition is still unknown, resulting in poor growth and development.

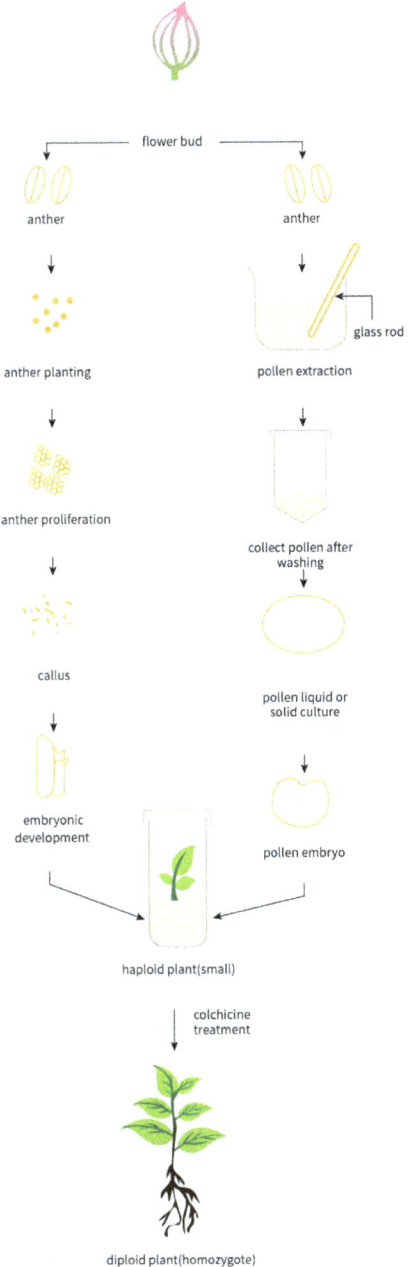

flower bud

anther

anther

anther planting

pollen extraction

glass rod

anther proliferation

collect pollen after washing

callus

pollen liquid or solid culture

embryonic development

pollen embryo

haploid plant(small)

colchicine treatment

diploid plant(homozygote)

Pollen and anther culture developed into haploid and diploid plants

In order to overcome these shortcomings, in 1977, Sunderland and other scientists improved the culture strategy and collected pollens by the natural dispersion method. After cold treatment at 7°C for two weeks, the flower buds or young spikes were allowed to grow on the surface of appropriate liquid medium. After the anthers naturally cracked and scattered pollens, they were centrifuged to collect pollens. The pollens were cultivated in the medium containing inositol and glutamine. The efficiency of pollen development in plants was improved.

Because pure pollen culture is not easy to achieve, some scientists still like to choose anthers as culture material, although anther culture sometimes does not produce haploid plants. This is because the anther is composed of the anther wall and pollen sac. The cells of the anther wall are diploid and can form a callus to develop into a plant.

Anther culture technology was successful in the 1960s, and it is still widely used in crop breeding. In 1964, two Indian scientists first bred haploid plants from the anthers of *Datura innoxia*, and proved that these small plants were derived from pollens. In 1982, Lichter used isolated microspore (pollen) culture of Brassica napus to induce embryogenesis and plant regeneration. In 2013, pepper pollen breeding was successful. In 2014, Chinese cabbage, cauliflower, and flax pollen breeding was successful. In 2019, the anther culture of the blueberry variety "summit" was successful [5]. In 2020, the anther culture of *Anthurium andraeanum* was also successful [6]. Through this technique, haploid plants of hundreds of plants have been obtained.

At present, more than 80 kinds of rice and 20 kinds of wheat have been obtained through pollen haploid breeding. China has also made a breakthrough in maize, which is generally considered to be more difficult and has a low frequency of haploid induction in the world, and has obtained more than 100 haploid plants.

The N6 medium and potato medium developed by China have received good haploid induction effect on rice, wheat, and other crops. They are not only widely used in China but also well received worldwide. In woody plants, China has obtained pollen plants of the main apple varieties, which is in a leading position in the world. In addition, haploid pollen plants of more than 20 woody plants, such as *Hevea brasiliensis* and *Populus nigra*, have also been cultivated.

Partial plant pollen culture

Plant	Pollen Development Stage	Basic Medium	Pretreatment Method	Regeneration Method	Research Year
Strawberry	Late uninucleate stage	NLN	Low temperature	Embryoid pathway	2011 (Meng Wang)
Calla lily	Mid-late uninucleate stage and early binucleate stage	NLN	High temperature	Callus pathway	2011 (Wang et al.)
Indonesian olives	Late uninucleate stage to early binucleate stage	NLN	Heat shock	Embryoid pathway	2011 (Winarto et al.)
Broccoli	Late uninucleate stage	NLN	Heat shock and dark	Embryoid pathway	2011 (NaHy et al.)
Roselle	Uninucleate stage	MS	Low temperature/high temperature/dark	Callus pathway	2012 (Ma'arup)
Chili	Initial binucleate stage/Late uninucleate stage to early binucleate stage	NLNS	Heat shock/dark/ mannitol	Embryoid pathway	2013 (Kim)
Ethiopian mustard	Late uninucleate stage	NLN	Heat shock and dark	Embryoid pathway	2013 (Yazdi et al.)
Chinese cabbage		NLN	Low temperature	Embryoid pathway	2014 (Liu Shi)
Cauliflower	Late uninucleate to initial binucleate period	NLN	Low temperature	Embryoid pathway	2014 (Gu et al.)
Flax	Mid-late uninucleate stage	HA		Callus pathway	2014 (Shumin et al.)
Brassica parachinensis Bailey	Late uninucleate stage	MS	Dark	Callus pathway	2015 (Yanchun Qiao, et al.)
Blueberry	Late uninucleate stage	MS	Low temperature	Callus pathway	2019 (Jin et al.)
Anthurium andraeanum	Later uninucleate stage	N6	Low temperature	Callus pathway	2020 (Danqing et al.)

4. Plant Cells for the Production of Traditional Chinese Medicines

There are about 300,000 species of plants in the world, including more than 30,000 species of higher plants. Plants are one of the important sources of food and medicine for human survival, and they contain tens of thousands of chemical compounds in their cells.

The story of Shizhen Li's compilation of *Compendium of Materia Medica* is well known to the world. In this masterpiece, 1892 kinds of drugs are listed, an overwhelming majority of which are plants. Plants usually produce some byproducts in the process of growth, that is, secondary metabolites. Many of the secondary metabolites are important drugs, which can be used to strengthen the human body.

The medicinal components in plant cells mainly include two major categories. One is cell inclusions, which are the storage or waste of cells, including alkaloids, such as ephedrine, caffeine, atropine, quinine, and berberine; glucosides, such as flavonoid glycosides, digitoxin, anthraquinone glycoside, and Shikonin; volatile oils, such as peppermint oil, clove oil, and eucalyptus oil; and organic acids, such as malic acid, citric acid, salicylic acid, and tartaric acid. The other is called physiologically active substances, including enzymes, vitamins, plant hormones, antibiotics, and plantfungicidin.

According to conservative estimates, more than 20,000 kinds of plant natural metabolites have been discovered, and the rate of newly discovered ones is increasing by 1,600 species every year. The byproducts produced by these newly discovered plants in the process of life activity may become new drugs. Up to now, 25% of legal drugs come from plants.

As the population increases, the demand for drugs is growing. Due to the influence of overexploitation and natural disasters, the resources of wild medicinal plants are increasingly exhausted. At present, more than 100 kinds of traditional Chinese medicines are extremely scarce. In the past, many precious medicinal herbs, such as *Gastrodia elata*, ginseng, *Angelica*, *Astragalus membranaceus*, poppy, and marijuana, were cultivated artificially. Due to the slow growth of plants, even large-scale cultivation cannot meet the actual needs.

So, what should we do? We know that plants are made up of cells, and their pharmaceutical ingredients are also produced by cells in the process of life. By large-scale cultivation of plant cells, the purpose of producing natural Chinese herbal medicines can be achieved. At this time, each cell becomes a micro pharmaceutical production factory. Plant cell culture is carried out in the indoor controlled environment, without any external impact of seasons and natural disasters.

In 1968, Reinhard and others pioneered the use of plant cell culture to produce drugs, and produced Harmine. Later, they produced diosgenin, ginsenoside, and visnadin. At present, the scale of cell culture tanks used for large-scale production of tobacco has reached 20,000 liters (20 tons). The size of the anti-tumor drug paclitaxel produced by taxus cells has reached the world advanced level of 60 milligrams per liter.

There are two main methods for industrialized cultivation of plant cells, one is suspension culture and the other is fixed culture.

Suspension culture is suitable for rapid proliferation of a large number of cells. In 1953, Muir successfully cultured the calluses of tobacco and standing marigold in suspension culture. Six years after that, Tulecke and Nickell invented a 20-liter enclosed suspension culture system for plants. The system consists of a culture tank and four auxiliary pipes. After sterilization by high-pressure steam, cells and medium to be cultured are added, and then stirred with compressed air without bacteria. After a period of culture, the nutrients in culture medium are exhausted, and the cells no longer proliferate and grow. At this time, the culture tank is opened to harvest the cells and extract the metabolites.

The advantages of this batch culture method are simple and not easily contaminated with bacteria, but the production efficiency is very low, and the accumulation of secondary metabolites is very small, so some people have improved the method on this basis. It is well known that cells in the metabolism process will produce some chemicals, such as lactic acid and ammonia; when these substances accumulate to a certain extent in the culture medium, it will inhibit the growth of cells. So, scientists came up with a brilliant idea; while the nutrients in the medium are consumed to a certain extent, a portion of the medium is started to be discharged and in the meanwhile, adding the same amount of fresh medium, this not only supplements the nutrients in the culture tank but also dilutes the toxic chemicals produced by the cells during the growth process, so that the cells start to grow vigorously again. After a period of cultivation, with the consumption of nutrients and accumulation of toxic substances in medium, fresh medium is added again. This culture method is called continuous culture or perfusion culture. Its advantages are obvious and can greatly increase the production efficiency.

Although perfusion culture increased the yield of cells, it was not conducive to the accumulation of drugs, so scientists invented the segmentation culture method. In the early stage of culture, nutrition was supplemented and oxygen supply was increased to promote the growth of cells. In the later stage, the nutrition and oxygen were no longer supplemented. At this time, due to the changes of living environments, the metabolic mode of cells is also forced to change, and the cells begin to accumulate higher concentrations of drugs.

Compared with suspension culture, fixed culture is more conducive to drug accumulation. The so-called fixed culture is to embed the cells inside the inert supports, or attach the cells to the surface of the inert supports. In 1979, Brodelius took the lead in the cultivation of *Morinda citrifolia*, *Catharanthus roseus*, and *Digitalis lanata* cells with calcium alginate as support. In the experiment, he found that the fixed cells tend to differentiate and form tissues, and such cells are more conducive to drug synthesis.

The common plant cell fixed culture methods include flat bed culture and column culture. First of all, in the flat bed culture method, the whole culture system consists of a culture bed, liquid storage tank, peristaltic pumps, and other parts. The bottom of the bed is a sterile flat pad woven with materials such as polypropylene, and the fresh cells are fixed on the flat pad. The sterile medium is fixed on top of the culture bed, and the medium is dripped down through the pipe to supply cells with nutrients. The culture medium consumed in the culture bed is sent back to the liquid storage tank by peristaltic pump. Although the equipment of this system is simple, it can synthesize natural drugs more effectively than the suspension culture method.

Another fixed culture method is column culture. It is a mixture of the cultured plant cells, agar, and sodium alginate to make one-by-one 1–2-centimeter-square cell masses that are concentrated in an aseptic column. In this way, the nutrient solution under the storage tank can flow through most of the cells, that is, the proportion of drip area is greatly increased, the synthesis of secondary biomass is greatly increased, and the floor area is greatly reduced. In column culture, because the cells are fixed, we should choose the seed cells that can synthesize drugs in cells, but secrete into the extracellular medium naturally or after induction. In addition, it should be noted that proper ventilation and illumination are beneficial to the synthesis of natural drugs in most cases.

Plant cell culture produces products of no more than two kinds, one is the cells themselves and the other is cell metabolites. The former includes cell culture of ginseng, lithospermum, and tobacco. At present, the scale of ginseng cell culture has reached 2 cubic meters, and the output of 10 kilograms of wet cells per day has also been reached in China. The wet cells are harvested and lyophilized to obtain active ginseng cell powder, which is both health food raw material and valuable medicinal material. The cultivating scale of lithospermum cells is also up to 750 liters, and the obtained lithospermum cells can be directly used for the manufacture of oral or topical anti-inflammatory agents, and can also be used for extracting shikonin. In Japan, tobacco cells were cultured by the two-stage method, and then collected as cigarette raw materials, with a scale of 20 cubic meters.

There are more than 50 major categories of primary and secondary cell metabolites produced by plant cell culture, including drugs, spices, oils, latexes, vitamins, pigments, hormones, polysaccharides, plant insecticides, and growth hormones, while the drugs include antibiotics, insulin-like polypeptides, and antitumor preparation. The content of more than 30 kinds of drugs in artificial culture has reached or exceeded the level of parent plants. For example, in cultured ginseng cells, the content of ginsenosides is 5.7 times higher than that of natural plants, and the cultured ginseng cells contain enzymes and other active components that are not found in natural ginseng and the health care effects are better than natural ginseng. The content of anthraquinone in the cultured *Spatholobus suberectus* cells was 8 times higher than that of the natural plant. In more than 200 plant cell cultures that have been studied, it has been found that more than 30 components which are useful to humans can be produced, many of which are important drugs that are widely used clinically.

Plant cell culture has greatly alleviated the shortage of precious medicinal plants and enriched the unique treasure house of Chinese herbal medicines. The search for new physiologically active components from natural products and development of new drugs has also become a global research hotspot.

Some drugs produced by plant cell culture

Compound	Cell Source	Effect
Digoxin	*Digitalis lanata* cells	Cardiotonic drug
Digitalis toxin	*Digitalis* cells	Cardiotonic drug
Reserpine	*Rauvolfia verticillate* cells	Antihypertensive drug
Quinine	*Cinchona* tree cells	Malaria drug
Vinblastine	*Catharanthus roseus* cells	Leukemia drug
Morphine	*Poppy* cells	Painkiller
Tetrahydrocannabinol	*Marijuana* cells	Psychotropic drug
Shikonin	*Lithospermum* cells	Anti-inflammatory drug
Ginsenoside	*Ginseng* cells	Health care products
Berberine	*Coptischinensis* cells	Antidiarrheal drug
Insulin-like polypeptide	Bitter gourd cells	Insulin-like polypeptide
β-glucocerebrosidase	Carrot cells	Gaucher disease (glucocerebrosidase deficiency)

5. Large-Scale Culture of Animal Cells

(A)   (B)

Bioreactor for animal cell culture

(A) Glass culture tank (2 liters); (B) stainless steel culture tank (20 liters).

There are some extremely important substances in human body fluids. They are expensive, but very effective in clinic. For example, urokinase is an important drug for the treatment of cardiovascular and cerebrovascular diseases. Interferon is an important drug against viral invasion. Erythropoietin is an effective drug for the treatment of pernicious anemia. They are rarely contained in the human body and are difficult to extract by conventional methods. Scientists have found that these drugs are proteinaceous and synthesized under the guidance of genes. If these drug genes are isolated and cloned, and then the drugs are synthesized in large quantities in ribosomes, the protein synthesis machines in cells, enough drugs can be obtained for clinical use. In fact, this is also an important direction for bioengineering applications.

In the practice of bioengineering, priority is often given to transferring drug genes into *Escherichia colis*. This is because they divide every 20 minutes, and the culture medium is relatively inexpensive, which greatly reduces the production cost. However, when some drug genes, such as erythropoietin (EPO), prourokinase (Pro-UK), and tissue-type plasminogen activator (t-PA), are transferred into *Escherichia colis*, the pharmaceutical proteins produced by culture are not active. Scientists have to consider transferring the drug genes into the animal cells (usually Chinese hamster ovary cells), and sure enough, the produced drugs have activities.

Further analysis reveals that the normal molecules of erythropoietin, prourokinase, and tissue-type plasminogen activator have a saccharide component. When these genes are expressed in *Escherichia colis*, the produced poietins have no saccharide components and no activities. When the genes are expressed in animal cells, the endoplasmic reticulums and the Golgi apparatuses in the cells can add saccharide chains to the newly formed drug molecules, thus showing their activities. Although the animal cells grow slowly and the medium is very expensive, scientists have not hesitated to choose animal cells because of the need for highly effective protein drugs.

In fact, the engineered cells that the drug genes have been transferred into have become a mini drug factory. However, in order to produce drugs in large quantities, there must be more pharmaceutical factories, that is, to cultivate engineered cells in large quantities, so that they can continuously secrete drugs. In general, people are used to referring to the large-scale cultivation of engineered cells (including animals, plants, and microbial cells) belonging to the category of cell engineering as the middle reaches of bioengineering, refer to the engineered cells with high drug production by genetic engineering as the upper reaches of bioengineering, and refer to the separation and extraction of drugs from the cultures as the lower reaches of bioengineering. Therefore, the massive culture of engineered cells plays an important role in the production of genetically engineered drugs.

There are two strategies for large-scale cultivation of animal cells: one is to expand the volume of the culture tanks and the other is to increase the culture density of the engineered cells.

The containers for culturing animal cells are called cell culture tanks, which are generally 2 liters or 5 liters used in the laboratories, and the tank body is made of heat-resistant glass. The industrialized cell culture tanks are generally much larger than these. The tanks are made of stainless steel and can be autoclaved *in situ*. Regardless of the types of culture tanks, they usually all have drain pipes, snorkels, and some instrument probes for detecting the environmental conditions in the culture tanks, such as thermometers, pH meters, and dissolved oxygen meters. At present, cell culture tanks are mainly manufactured in China, America, Germany, Japan, and other countries.

In industrial production, the volume of cell culture tanks cannot be expanded indefinitely, as this will greatly increase the workshop area of the factory and the risk of being contaminated. Once the cultured cells are contaminated, it will cause huge economic losses. Therefore, compared with expanding the volume of culture vessels, increasing the culture density of engineered cells is obviously more cost-effective in economy.

There are many ways to increase the culture density of animal cells. The earliest and most commonly used method currently is the microcarrier method, which was invented in 1967 by Dutch scientist Van Wezel. The so-called microcarriers are actually small balls that can be clearly distinguished under a microscope, usually made of gelatin, plastic, glass, cellulose, dextran, etc. Cells can be adhered to the surface of the microcarriers to grow. Since the balls have the largest surface area to volume ratio, more cells can be adhered to them to grow, thereby increasing the density of cells per unit volume. In the case of cultivation in small quantities in the laboratory, in order to save costs, a spinner bottle with a branch sampling port on one side can be used. It is made of heat-resistant glass, and used in combination with a magnetic stirrer. It forms a series from 100 milliliters to 1000 milliliters, and can be selected according to different needs. Prior to cultivating, animal cells and microcarriers were added to the spinner bottle under sterile conditions, and then transferred to a constant temperature incubator. Adjust the rotation speed of the magnetic stirrer, and let it drive the magnetic stirring rod in the spinner bottle to rotate so that the microcarriers and cultured cells are suspended, which is conducive to the full contact of cells with nutrients and the attachment of cells on microcarriers. After a period of culture, the pink medium turns pale yellow, which is due to the change of pH value of the medium caused by the lactic acid produced by the cells during their lives. At this time, it is necessary to replace the consumed medium with fresh medium in time.

If it is to explore a new production process, it is best to use a computer-controlled culture tank. Start with a small cell culture tank, such as a 2-liter glass tank. The body of the cell culture tank is inserted with a thermometer, pH meter, dissolved oxygen meter, liquid inlet pipe, liquid outlet pipe, sampling pipe, air inlet pipe, etc. In the specific culture, after the tank body and its attached pipes and probes are sealed with cotton cloth and kraft paper, high-pressure steam disinfection is carried out first, then the microcarriers and seed cells are added into the tank under sterile conditions, and then the tank is installed on the base and connected with a control panel. Adjust the temperature, pH value, dissolved oxygen, stirring speed, and other parameters on the control panel to make the cells grow in the appropriate environment.

During the cell culture process, it is necessary to periodically sample and analyze various parameters such as glucose consumption, amino acid consumption, lactic acid accumulation, ammonia accumulation, cell density, and numbers of drugs produced by the transferred genes. Then, using these parameters as the ordinate, the cultured days as the abscissa, some curves are drawn, which is convenient for analyzing the culture technology. As for whether the cells are evenly distributed on the microcarriers and whether their morphologies are normal, samples can be taken and observed clearly under an inverted microscope. When the culture technology is mature, the scale of culture can be expanded gradually until it is suitable for industrial production.

Generally, the cell density of microcarrier culture is 1–2 million cells per milliliter. However, if microcarrier technology and perfusion technology are combined, cell density can be increased by an order of magnitude, thus greatly improving the production efficiency. The so-called perfusion culture is that during the culture process, the peristaltic pumps controlled by the computer continuously extract the consumed medium, and at the same time, fresh medium is added, so that the cells can vigorously divide and grow in a good nutritional environment. At present, the cell density of perfusion culture of nearly 100 million cells per milliliter can be reached, but considering the density of cells in the human body is 2–3 billion cells per cubic centimeter, this technology still has great potential for further development.

The perfusion culture technology is highly automatic, which reduces the risk of being contaminated, and is particularly suitable for industrial production. The engineered cells secrete protein drugs into medium, and the drugs can be isolated and purified by continuously collecting the extracted medium.

When cultured with microcarriers, animal cells can only grow on the surfaces of the microcarriers, but for some genetically engineered cells, their growth is suspended, and they cannot be adhered to microcarriers. On the basis of microcarriers, porous microcarriers, also known as porous microspheres, have been developed. The preparation materials mainly include gelatin, collagen, glass, plastic, cellulose, ceramics, and sodium alginate. The manufacturing process is far more complicated than microcarriers, mainly because the size and pore diameter of porous microspheres are not easy to be controlled; however, a variety of porous microspheres have been successfully manufactured in China. The size of porous microspheres is similar to that of microcarriers, but the surface of porous microspheres has many pores, and its interior is almost empty. The cells can grow both inside and on the surface of the porous microspheres, thereby greatly increasing the cell density. Hybridoma cells can be cultured at a density of 50 million cells per milliliter of culture medium, while adherent genetically engineered cells can even reach 1–2 billion cells per milliliter of microspheres, greatly reducing the cost of industrial production. Griffith, a famous biologist in England, highly praised it as the second stage of the development of microcarriers and a revolution in cell culture technology.

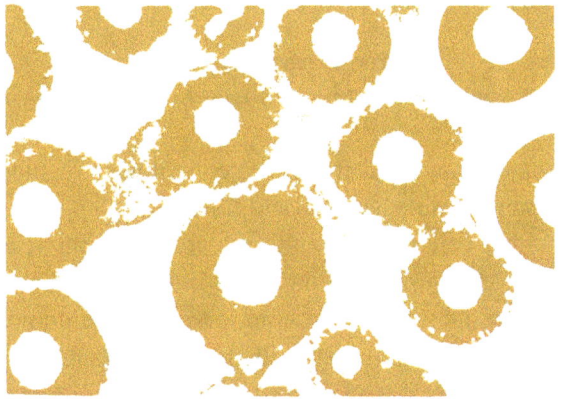

Animal cells can grow in multiple layers on microcarriers

In addition to being used for the massive culture of animal cells to produce biological products, some biodegradable porous microspheres can also be used as a new therapeutic method, that is, to cultivate bone marrow and blood cells with important medical values by using porous biomaterials. Because the cells can grow stereoscopically in the porous network, they can be designed as an *in vitro* apparatus for supplementing metabolic activity of dysfunctional organs or for *in vivo* transplantation of the cultured cells.

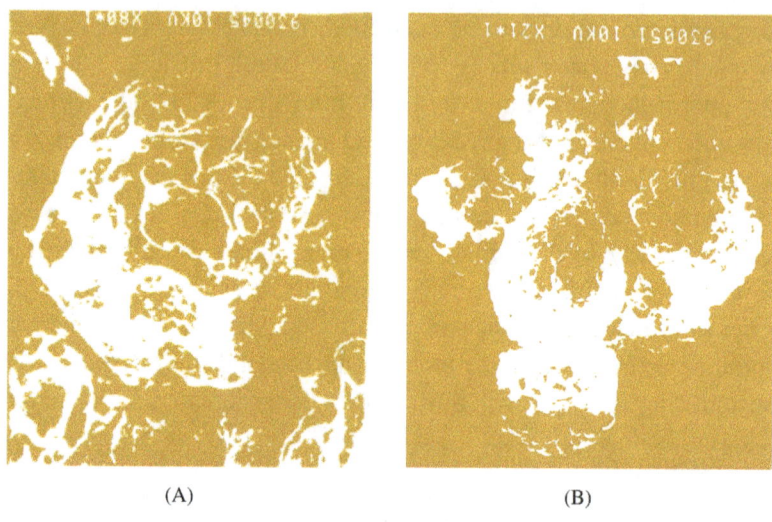

(A)   (B)

Porous microspheres

(A) Electron micrograph, before cell culture; (B) electron micrograph, after cell culture.

The more mature animal cell culture technologies include the hollow fiber method, microcapsule method, and fluidized bed method. At present, the scale of animal cell massive culture technology has reached several thousand liters or even up to ten thousand liters. Over the past few decades, many cell products with important practical and commercial value have been produced using animal cell massive culture technology. For example, in addition to the aforementioned genetic engineering drugs, there are tumor necrosis factor, interferons (α, β, γ), tissue plasminogen activator, growth hormone, serum protein, etc. Another example includes various virus vaccines, because the viruses can only survive and reproduce within living cells, and thus the production of viral vaccines cannot do without living cells; a large number of vaccines produced by cell culture technology, including foot and mouth disease vaccine, rabies vaccine, polio vaccine, bovine leukemia vaccine, and measles vaccine, are all produced under strictly controlled conditions. Others include monoclonal antibodies, which not only have a great variety of types but also a wide range of applications. Moreover, some cultured living cells can also be used as therapeutic agents.

In recent years, with the development and improvement of gene technology and cell culture technology, a kind of living cell therapy has emerged in the world, which uses living cells as a therapeutic agent to treat various difficult genetic diseases, including cancer. This emerging medical technology mainly uses the genetic engineering method to propagate the patient's autologous cells *in vitro*, including lymphocytes, bone marrow cells, tumor infiltrating cells, thymocytes, and other living cells, so as to expand or produce substances with therapeutic effects, such as antibodies, proteins, and hormones, and then inject or implant these living cells into the patient's body to treat diseases such as malignant tumors and blood cancer. From the clinical experiments and applications all over the world, this live cell therapy has obvious curative effects on serious genetic diseases and infectious diseases such as cancer, leukemia, diabetes, hemophilia, burn, and AIDS. Taking cancer treatment as an example, its biggest advantage is that it can attack the spreading cancer cells without harming normal cells.

After long-term development, animal cell massive culture technology has become more and more mature. It is an important field of cell engineering, and its development prospect is very attractive.

6. Artificial Skins, More Beautiful

Among all the organs of the human body, except the liver and brain, no other organ has more complex and diverse functions than the skin [7]. It covers the whole body, and is an important gateway of the human body, but also the first line of defense against the invasion of various external harmful factors. The epidermis, dermis, and subcutaneous tissue in the skin form a unique three-dimensional defense line, which can eliminate or reduce the damage to the human body caused by various external physical factors (such as friction, extrusion, traction, high temperature, low temperature, radiation, and ultraviolet radiation), chemical factors (such as acid, alkali, cosmetics, and topical drugs), and biological factors (such as bacteria, viruses, fungi, and parasites).

Not only that, the skin also has absorption functions. Normal skin rarely absorbs materials like gas, water, and electrolytes, but fat-soluble substances, oils and fats, heavy metals, salt substances, inorganic acids, and the like can be absorbed by the skin to varying degrees. The absorption function of the skin is also the theoretical basis of skin medication. The skin also has a secretion function. The sebaceous glands secrete and excrete sebum, which can lubricate and protect the skin and hair. This function of the skin can make a person have a head of beautiful hair, with a healthy glow all over the face. The skin can also participate in the manufacture of vitamin D3, regulate the body's water and salt, store blood and fat, and play an important role in the human blood bank and energy bank. Of course, the skin does more than that. To say the least, even if the other functions of the skin are normal, from the aesthetic point of view, disfigurement due to skin conditions will make people feel depressed.

It is precisely because the skin is the gateway of the human body that it is easy to be injured or infected with diseases. For example, large area burns, scald, skin suppuration, erosion, ulcer, and various skin diseases caused by microbial invasion often need skin transplantation in the treatment process. If the skin of another person is transplanted to the patient, there will be immune rejection, that is, the transplanted skin cannot grow on the patient. Taking the patient's own skin can avoid this problem. But, for the patient, this is undoubtedly worse. It is just like robbing Peter to pay Paul.

For a long time, scientists have successfully developed a variety of skin substitutes, which greatly enriched the clinical treatment methods and alleviated the pain of patients. Among them, a kind of "artificial skin" that appeared in places such as Shanghai City and Dalian City in recent years attracted people's attention. Although it is skin, to be exact, it is not the skin of flesh and blood, but a kind of medical gauze and wound dressing made of chitin fiber. Compared with the clinically applied pigskin covering material, it has many advantages and is inexpensive.

Chitin widely exists in the shells of shrimps, crabs, and insects, as well as in the cell walls of fungi and algae. It is non-toxic, degradable, and compatible with the human body. Artificial skin involves treating chitin in 12-micrometer-wide filaments, and then processing it into a variety of medical dressings. At present, in the laboratory, the annual output of 1 ton can be achieved. The advantage of using chitin as a skin substitute is that it has good air permeability and water permeability. After being applied on the surface of a burn, scald, ulcer, and bedsore, it can not only protect the wound from infection but also has a strong therapeutic function. It does not need to be replaced during treatment, and will fall off automatically after the wound is healed. The cytotoxicity test, hemolysis test, and skin primary stimulation test are also in line with the medical treatment requirement. But, artificial skin is just like the food "artificial meat" made of soybean proteins. After all, it is not the real skin and cannot perform some important functions of skin.

From the plant cells that can grow into the whole plant after culture, we are inspired. Can we culture skin cells to regenerate skin? In theory, of course, it can, but it is difficult to implement in technology, because animal cells are unlike plant cells that still have good totipotency even after differentiation. As for general animal cells, once differentiation occurs, totipotency is difficult to recover, but this is not absolute. Some skin cells will rejuvenate and continue to divide and proliferate after changes in external environment.

Since 1997, in America, Germany, Russia, Japan, Britain, and China, skins have been successfully prepared by using cell culture technology. Although they are also artificial, they are organic skins with dermal structure and function.

The organic skin prepared by Nankai University in China used a kind of natural polymer material as carrier scaffold so that the living natural skin cells can adhere, grow, divide, and proliferation on it. When the skin piece grows to a certain extent, it can be used in clinic. After skin transplantation, within a certain period of time, the carrier scaffold will degrade and produce new extracellular matrix, thus forming a new skin, which is similar to the purpose of self-skin transplantation. The skin prepared by cell culture is an active skin, equivalent to dermis. It can be used in skin transplantation for repairing after large-area skin injury, and can also be used in skin transplantation for patients who cannot provide their own skin graft (for example, diabetic patients with skin erosion). It is a kind of artificial organ product with wide application prospects.

Not long ago, scientists at the University of Tokyo invented a new technology that can recreate human beings. Although the creation of human beings is still in the science fiction stage, the new artificial organ technology has been put into practical use. These scientists have mainly developed a new type of substance that provides suitable conditions for tissue regeneration. Using this material to make a three-dimensional scaffold, human skin cells and other cells can grow on it, and self-organize into the necessary shape and structure, just as the human skeleton can support the human body. After the growth has finished, this scaffold gradually degrades, leaving only the grown tissue.

The scaffold technology itself is not novel, because many countries in the world are using it as the basis of artificial organs. But, the ingenuity of these Japanese scientists is that they combine two kinds of existing scaffolds (i.e., ordinary polymeric collagen and synthetic polymer mesh) together to avoid their respective shortcomings when used alone. Because collagen is very fragile, and the surface of synthetic polymer is not conducive to cell growth; when combined, they can learn from each other. In practice, they mix the polymer mesh and collagen sponge together, and the sponge forms a second net inside the polymer mesh to increase the strength of the overall structure. This method is simple and original, and no one has thought of it before. The strength of the composite is almost the same as that of the general polymer network, but it is more elastic than the use of the original gum sponge alone.

Although they can neither make eyes that can stretch and rotate nor create forearms that can hold things up, the technologies they have already mastered can be used to cultivate artificial skins, blood vessels, and even some of the most important living organs such as heart valves, bones, and lung tissues. Now, only artificial skins can be used in practice.

By culturing the prepuce cells of newborn penis, American scientists have successfully obtained another kind of dermal skin. Once, a hospital in the United States used the skin to perform transplants on 17 patients. The used skin was supplied by a pharmaceutical company in New Jersey and was very expensive. The 17 treated patients, including 15 children and 2 adults, suffered from a special condition characterized by hypersensitivity of the skin, which can cause blisters even with a slight touch. A couple from Colorado is happy with the results of their two-and-a-half-year-old daughter undergoing surgery. Scientists analyzed that the newborn cells have strong vitality and vigorous division, which may be the main reason for the success of the experiment.

Apligraf is a kind of artificial skin product with double-layer living cells produced by Organogenesis Holdings Inc., USA. It is also a relatively mature tissue engineering skin product. Its cell composition is allogeneic epithelial cells and fibroblasts, all from neonatal foreskin, and the transplantation success rate is high. It has been approved by the US Food and Drug Administration for clinical application in the treatment of chronic skin ulcer, diabetic foot, and other diseases.

The artificial skin made in China is named ActivSkin, which began to be massively produced in early 2008 and has been used in clinical

Tissue engineering skin products that have been approved or undergoing clinical trials

Product Name	Indications	Manufacturer
Alloderm	Burns, scalds	Life Cell, USA
Integra	Third-degree burns, scalds of large area	Integra Lifesciences Corporation, USA
Epicel	Burns, scalds	Genzyme Biosurgery, USA
TransCyte	Second-degree and third-degree burns	Advanced Tissue Science, USA
Apligraf	Chronic skin ulcer	Organogenesis, USA
Dermagraft	Chronic skin ulcer	Advanced Tissue Sciences Inc., USA
		Smith and Nephew, UK
EpiDex	Chronic skin ulcer	Euroderm, Germany
Epibase	Chronic skin ulcer	Laboratoires Genevrier, France
Myskin	Chronic skin ulcer	CellTran, UK
OrCel	Chronic skin ulcer	Ortec, USA
BioSeed-S	Chronic skin ulcer	BioTissue Technologies, Germany
Hyalograft3D Laserskin	Chronic skin ulcer	Fidia Advanced Biopolymers, Italy
ActivSkin	Deep second-degree burns, third-degree burns that are not exceeding 20 square centimeters (diameter less than 5 centimeters)	Shanxi Aierfu Tissue Engineering Co., Ltd., China

medicine. It took about 10 years before and after the research and development of ActivSkin, with 80 million yuan of investment in scientific research, while the research and development cycle of a similar product in the United States is 18 years, with an investment of 460 million US dollars.

The advantage of using allogeneic cells for skin transplantation is that patients can be treated in time, but there is also a risk of immune rejection. The real ideal skin transplantation should be to cultivate the skin cells of the patients themselves, and then use them for transplantation, which will greatly improve the survival rate of transplantation. However, it is not suitable for those patients who need urgent treatment, because the cells grow slowly in the culture medium and the growth cycle of the skin is too long.

But, for other kinds of patients, it is a great gospel. Their skin has normal physiological function, but it is very bad, which seriously affects its beauty, for example, very rough skin, large-area birthmark, large facial scar, pimple face, and pockmarked face. As the saying goes, "the love of beauty, everyone has it". These kinds of patients are very distressed, mentally enduring the anguish. If one can cultivate their own skin cells, and then transplant them, so that they have a beautiful face, it will undoubtedly make them feel very happy. With the development of science and technology, it can be done in the future.

7. Cultivating Organs, the Transplantation of Kidneys and Livers

For a machine, when an individual part is damaged, it will not work. When the damaged part is replaced, the machine will resume normal operation. In fact, the human body is also like this. From ancient times to now, as long as any vital organ is damaged or exhausted, it may lead to death of the whole human body. This vital organ can be either the kidney, or liver, or heart, or lung. If it is only kidney failure, and all other vital organs (heart, lung, liver, etc.) are intact, it will also lead to death.

So, can the human body be like a machine, when any part of it is damaged, we just replace it with a new part to make the machine work again without being scrapped? On May 30, 2000, scientists from the

Aktyubinsk State Medical School in western Kazakhstan developed new livers from embryonic stem cells. They injected embryonic stem cells into a mouse with cirrhosis and found that a second liver developed next to the diseased liver, which successfully replaced the original function of the diseased liver. A mouse that is suffering from this disease will usually die within a week or two. After this treatment, the sick mouse survived. The vice dean of the medical school, Istulioff, said that it is the first time that the mouse can get a new liver, an absolutely new healthy liver, and its function was exactly the same as the original. The treated mouse had been already able to run around, eat, and perform other activities. The experiment also showed that the injected cells did not damage the brain or other vital organs of the mouse, indicating that the treatment is safe and reliable.

Regeneration of mouse liver after injection of stem cells

Although this experiment is carried out on animals, it is also important for humans. If this method is applied to humans, it can save patients with liver cancer or cirrhosis, because these diseases require the replacement of the liver.

As early as the 300 BC, in ancient China, we had the magical legend of using organ transplants to treat diseases. There was a story about the famous doctor Que Bian of the Warring States Era in the book *Lie Zi*, describing the heart exchange between Zhao and Lu. Now, it is universally acknowledged that this is the earliest assumption of organ transplantation. After all, this is only a legend and an illusion. According to the medical level at that time, it was impossible to do such surgery.

The real organ transplant experiment was initiated in the early 20th century. So far, nearly one million people with incurable diseases have acquired a second life chance through organ transplantation such as kidney, liver, heart, and pancreas and bone marrow transplantation. The organs used for transplantation are expensive and have very few sources, which makes many patients wait to be treated. From January 1, 2015, China has completely stopped using executed criminals as a source of transplant donors; voluntary donation of organs after the death of citizens has become the only legal channel for organ transplantation.

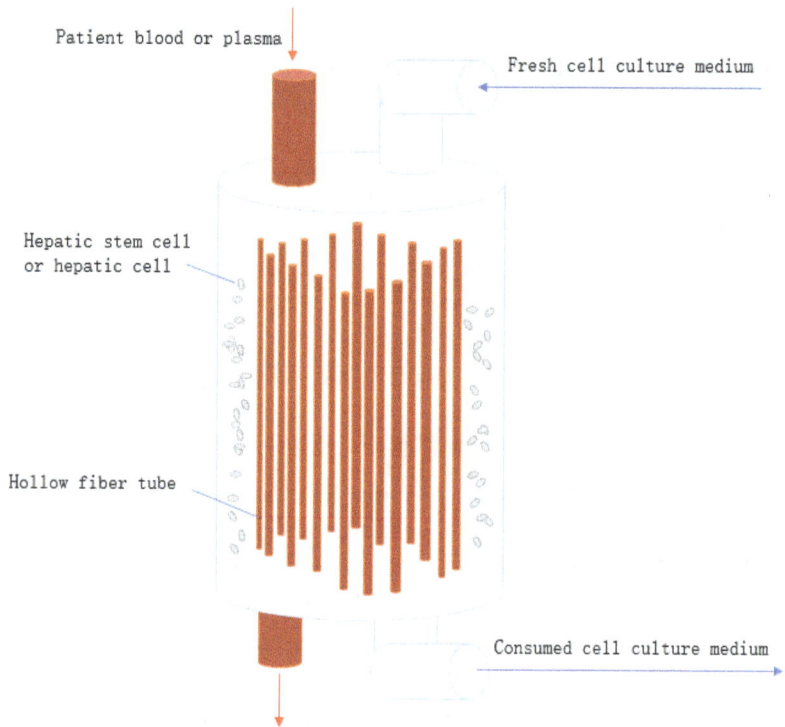

Bioartificial liver supporting system

The use of stem cell culture to produce artificial organs or living functional cell devices *in vitro* including bioartificial liver is expected to ease the shortage of organ supply and enable more patients to receive effective treatment. At present, research on the use of pluripotent or unipotent stem cells to cultivate human cells and tissues has achieved certain progress. However, the more promising application is the totipotent stem cell with the strongest differentiation ability, which can only be obtained through embryos. A fertilized egg formed a structure called blastocyst in the early stage of divisions before it is implanted into the uterus, which consists of about 140 cells. At one end of the interior of the blastocyst, there is an inner cell mass, which is a collection of embryonic stem cells with full differentiation capabilities. If it can be taken out to culture, different tissue cells and even organs can be induced *in vitro* for transplantation.

Before the advent of somatic cell cloning technology, scientists could only obtain stem cells from aborted, stillbirth, or artificially inseminated human embryos for research. The advent of the cloned sheep Dolly means that people can clone human embryos through somatic cells, which will make it easier to obtain stem cells. Doctors can take somatic cells from patients for cloning, and the formed blastocysts are developed for 6–7 days, and then the stem cells are extracted therefrom, and the cells, tissues, or organs whose genetic characteristics are completely matched with the patients are cultured [8], and then the cultured tissue or organ is transplanted to the patients who provide the cells, which is called "therapeutic cloning". Compared with other artificial organs, the superiority of therapeutic cloning is that no rejection occurs, and the success rate of surgery is high. Once this technology matures, blood cells, brain cells, bones, and internal organs can be replaced, bringing hope to patients with diseases such as leukemia, Parkinson's disease, heart disease, and cancer.

In Singapore, scientists have also acquired the organs needed for transplantation through other technical means. A few years ago, researchers from the Department of Engineering, Department of Biological Sciences, and Department of Medicine of the National University of Singapore collaborated to extract tissue cells from diseased organs, culture them in the laboratory, then inject them back into the original organ, and then let them naturally proliferate and regenerate the diseased organs in the body. Not long ago, they had cultured cartilage tissues for three patients, and the condition of patients has been very stable until now. Besides cartilage, the research group of the medical department has also carried out *in vitro* culture experiments of nerve, heart vessel, and liver in animals, and the results are encouraging.

Professor Yongxing Li from the Medical School of National University of Singapore told the media that in the future, when human organs are damaged or infected by germs, doctors will not need to transplant artificial organs for them, but will obtain tissue cells from the diseased organs, culture them in the laboratory, and they will be injected back into the original organs, allowing the cells to naturally proliferate and regenerate the diseased organs. Because of using its own tissue cells, the body will not reject them. He revealed that they will also try to extract eggs from women for culture. In theory, since the eggs are the cells of the human body, they should be able to successfully "fuse" and reproduce no matter whether they are transplanted to any organ. At the beginning of the 21st century, Professor Ruixing Zhang at the Major of Biomedical Engineering of Department of Engineering of National University of Singapore also told the media that his laboratory had successfully completed the cell culture of bone and skin in animals. He hoped to apply this technology to people in the next 5–10 years. However, there is no relevant report about the final results. In Russia, scientists transplanted the pancreas of mice into a diabetic child, trying to cure this difficult disease, and the result was a great success. This new method for the treatment of juvenile diabetes was developed by the Russia Institute of Transplantation and Artificial Organs, which carried on clinical trials at the Children's Hospital. The medical experts made shallow anesthesia of a new-born big-ear mouse, then took out its pancreas, put it into the nutrient solution for cultivation, and transplanted it into the abdominal muscle of the child two weeks later. The mouse pancreatic tissue transplanted into the human body will produce insulins and other essential substances for the human body. After transplantation of mouse pancreas, the condition of the child improved rapidly and the height began to increase. The pancreases of 40 mice are sufficient to treat a child with diabetes.

Children with diabetes often have the so-called Mauriac syndrome, which inhibits the development of various organs of their bodies, making the children less than 150 centimeters tall and resembling a dwarf. After the treatment with the new method, the condition of the sick children is not only greatly reduced but their height can be increased by about 20 centimeters within one year.

In the past, Russian scientists have obtained drugs for the treatment of childhood diabetes from pancreatic cells of human embryos with miscarriage in late pregnancy. Later, Russian medical experts tried many times to transfer animal pancreatic cells into the human body, and the results showed that the pancreas of rats with big ears was successfully transplanted into children for the first time in the world. This method has a variety of advantages: first, the mouse pancreas transplantation into children with diabetes does not produce rejection reaction, the treatment effect is ideal; second, patients only need to inject mouse pancreatic cells once a year, and the treatment method is simple, the patients have little pain, and the mouse pancreatic gland is easy to obtain.

In Japan, scientists have developed pigs (chimeric pigs) with human organs for organ transplantation. Pig organs, no matter in volume size or in terms of functional parameters, are very similar to human organs. As long as the rejection reaction that occurs during the transplant process can be overcome, the pig's organs can still be used by humans. Researchers at Nagoya University experimented with human DNA cells implanted into fertilized eggs in pigs. And, not long ago, they announced that they had developed a new "chimeric pig" variety for the first time in the world. They used modern biotechnology to fuse various enzyme genes in human blood with frozen embryos of pigs, and then implanted them into the uterus of 19 sows. As a result, three sows successfully gave birth to 27 piglets mixed with the human blood gene within the scheduled time. Under the careful nurture of scientific researchers, they have very robust growth and development. It is generally believed that the blood bridge between humans and pigs is an important step toward practical organ transplantation, which is conducive to the settlement of pig organs in the human body. This achievement has been acknowledged widely by the international biological community, and millions of lives will be saved.

Coincidentally, British researchers are also fond of pigs. British scientists injected human DNA into the embryo of a pig. Six months later, a transgenic pig was born on the operating table. Scientists hope that human genes can coordinate pig organs with the human immune system, thus solving the growing shortage of organs for organ transplantation. A British company has more than 200 members of its transgenic pig family. The company is doing further experiments to get human blood flowing through the hearts of transgenic pigs to see what happens. "It's absolutely certain that the heart of transgenic pigs is much better than that of normal pigs, and there are fewer signs of immune rejection," the scientists in charge of the study told the International Allogeneic Transplantation Conference in Washington. This can lead to increased survival of xenografts in patients.

Since then, the British company has started human trials. Two other American biotechnology companies are doing the same work, and have transplanted tissues from genetically modified pigs to primates. This technology will not only be the progress of medicine but also has considerable economic value. In human organ transplantation, the viscera of primate apes are generally favored by people. But, more and more medical studies have found that some viruses in monkeys are very harmful to human beings. So, the medical experts turned their eyes to pigs, hoping that pigs could become the biggest donor of human organ transplantation.

Scientists in the United States have gone even further in artificial organs. They have tried to design and produce artificial organs "from scratch" through gene synthesis technology, and hope that this kind of artificial organ will be born within two years.

In fact, a gene is a small piece of deoxyribonucleic acid (DNA). It is the molecule that makes up the blueprint of life, made up of twisted double strands, each containing various quantity of bases. Some gene molecules can contain millions of base pairs. For a long time, scientists found it difficult to connect more than 100 base pairs together after numerous experiments. The research team led by Professor Wensi Yi, director of the Center of Genetic Sciences and Technologies of University of Texas, created short strands of DNA and then linked them together, enough to form the basis of a simple life.

Professor Wensi Yi said that they have created DNA that follows the structure of human tissues, which will enable humans to create artificial organs. He plans to copy key genes from each cell, and then select the best genes to link them together so that the artificial organs can survive and become the most effective organs.

The combination of tissue engineering technology and 3D printing technology provides a new opportunity for tissue regeneration. On the one hand, various cells and biological materials can be used to print out the rudiment of certain tissues and organs *in vitro*, thereby cultivating functional tissues and organs for clinical transplantation applications; on the other hand, cells and degradable biomaterials can be used as "ink" for *in situ* printing defect tissues and organs *in vivo*, which saves the steps and costs of transplanting tissues and organs cultured *in vitro*. The shape of the rudiment of the tissues and organs printed in 3D is fixed, but since there are living cells in the tissues and organs, and the cells are growing, the shape of the tissues and organs changes with the passage of time. This printing technology, which can change with time, is called 4D printing, and has important application prospects in the field of tissues and organs regeneration. As for 5D printing, the printed object can not only change its shape over time but also the function. This 5D printing concept is very suitable for the production of tissues and organs, because the shape of human tissues and organs can be directly printed, but the physiological functions cannot be printed directly, and can only be realized with the growth and development of cells in the future. Unfortunately, so far, regardless of 3D, 4D, or 5D printing, it has not created a tissue or organ with truly physiological functions, especially the more difficult printing of blood vessels and nerves with physiological functions.

The artificial tissues or organs with physiological functions may be realized in 5D printing in the future. At that time, transplanting liver or kidney may be as simple as changing parts of cars.

Chapter 3

The Profound Mysteries of Cell Fusions

1. The Hybrid Cells for Monoclonal Antibody Production

Cell fusion is one of the earliest developed cell engineering techniques. Many miracles can be created by fusion between cells of the same type or different types, among which monoclonal antibody (McAb) is a good example. So, what is a monoclonal antibody? How is it made?

We are not unfamiliar with immunity. It is an important function of some organisms.

Relying on this function, the organisms can distinguish their own components from non-self components to destroy and reject the bacteria, viruses, and other foreign matters that invade the body, and even the tumor cells that are transformed from the mutated normal cells *in vivo*, thus protecting the organism. It can be said that the immune system evolved in humans and higher animals is a veritable patron saint of the human or higher animal health. When the function of the immune system is reduced or destroyed, the organism will be very susceptible to diseases. For example, the frightening HIV viruses specifically destroy the human immune system.

The immune system mainly consists of lymphocytes, including T lymphocytes (also known as thymus lymphocytes) and B lymphocytes (also known as bone marrow lymphocytes).

When antigens (such as bacteria, viruses) invade higher organisms, on the one hand, T lymphocytes will produce a variety of lymphokine to repel antigens, making them difficult to function. On the other hand, with the help of T lymphocytes, B lymphocytes differentiate to produce many plasma cells. Each plasma cell is like an arsenal, which can produce countless weapons for killing bacteria, viruses, or cancer cells. Those weapons are called antibodies.

The structure of the antibody corresponds to the structure of the antigenic substance that induces the production of antibodies. The antibody and its antigen can be combined with each other like bolt and nut. When the antigen is wrapped by the antibody, the action of the antigen is bound and it is unable to cause trouble in the body. Eventually, the phagocytes will devour and digest the trapped antigen, and the invaded organism will be safe.

Newborns can get antibodies by sucking their mother's milk, so they are not easily infected with diseases; children can obtain antibodies through vaccination, and have immunity to some diseases. When small-pox, a severe infectious disease, was not extinct, young people had to vaccinate with a vaccine called vaccinia. Vaccinia is originally a disease of cattle. Because of the similar symptoms with human smallpox, people extract vaccinia viruses and make live vaccines after reducing the toxicity of viruses. When the attenuated vaccinia viruses, or antigens, are introduced into the human body, the human immune system acts immediately and produces antibodies against vaccinia viruses. After this kind of training, some day, when smallpox viruses really invade, B lymphocytes will produce antibodies more quickly, encircle the invading viruses, and wait for phagocytes to phagocytize and digest, so as to prevent people from suffering from smallpox. For another example, we all know that people who have suffered from a certain disease, such as poliomyelitis, will not get it the second time. This is because antibodies are produced in the body under the induction of pathogenic antigens. These antibodies have a good memory. When the enemies invade for the second time, the antibodies act as weapons to eliminate them.

It is precisely because antibodies can be combined with pathogenic microorganisms that invade organisms that they can be used clinically to treat some diseases. In the beginning, it is necessary to extract the antigens first, inject them into the animal, produce the antibodies, and then extract the desired antibodies from the serum. The number of antibodies prepared by this method is extremely limited, which limits their clinical application. Moreover, antibodies extracted from the serum are actually a mixture of various types of antibodies. This mixture is known as polyclonal antibodies, which can deal with the invasion of many foreign pathogens, but the probability of simultaneous invasion of multiple pathogens is quite rare. In this way, in practical applications, polyclonal antibodies appear to be both wasteful and inefficient.

Scientists think that if a lymphocyte that makes particular antibodies can be cloned to make monoclonal antibodies against certain specific pathogenic microorganisms, the monoclonal antibodies will become powerful weapons against diseases. Fortunately, it is not difficult to isolate B lymphocytes producing monoclonal antibodies, but the problem is that some cells derived from B lymphocytes have a very short survival time and can only proliferate for a few generations. In this way, we still could not obtain enough monoclonal antibodies, so there is no practical application value. Is there any way to make B lymphocytes immortal?

Scientists have noticed that the tumor cells cultured *in vitro* are almost unlimited and can reproduce indefinitely. Therefore, some people think that tumor cells can be isolated, and injected into other animals to produce antibodies, and then the antibodies are separated out to obtain monoclonal antibodies. But, unfortunately, this meaningful exploration did not achieve the desired results, and the prepared antibodies could neither react to the antigen nor be specific. The experiment ended in failure.

But, scientists did not give up exploration. Soon, some people observed in the experiment that some different kinds of animal cells can fuse with each other, which greatly opened the minds of scientists. Since B lymphocytes are capable of producing antibodies, and tumor cells can proliferate indefinitely. If these two cells are fused together, is it possible to obtain cells with new functions that can produce antibodies and proliferate indefinitely? This assumption paved the way for a final solution to the production of monoclonal antibodies.

In 1975, only two years after the success of DNA recombination technology, biotechnology made another key breakthrough. Köhler and Milstein, two scientists from the Molecular Biology Laboratory of University of Cambridge, UK, had successfully fused B lymphocyte that can secrete single antibodies with a myeloma cell that can proliferate indefinitely to artificially create a hybrid cell and clone it.

This hybridoma cell inherits the genetic characteristics of the two parental cells, which not only preserves the ability of rapid proliferation and passage of myeloma cells *in vitro* but also inherits the ability of B lymphocytes to synthesize and secrete antibodies, and the hybridoma cell can be used to massively produce monoclonal antibodies. For this invention, Köhler and Milstein won the 1984 Nobel Prize in Physiology or Medicine.

The hybridoma cells are prepared as follows: First, the mouse is immunized, that is, an antigenic substance is injected into the mouse, and then the spleen cells are separated from the lymphoid organ spleen of the mouse and fused with the myeloma cells, and then the hybridoma cells capable of producing the desired monoclonal antibodies are selected.

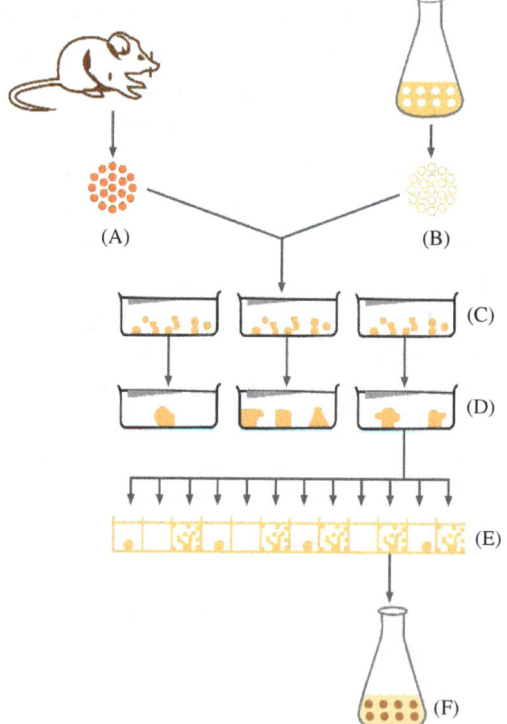

Process of monoclonal antibody preparation

(A) Lymphocytes; (B) myeloma cells; (C) cells growing in porous culture plates after fusion; (D) hybridoma cells growing while other cells died; (E) the specific antibodies in the culture medium are detected and the positive cells secreting the specific antibodies are cloned; (F) monoclonal antibodies are prepared in large quantities.

Since hybridoma cells are quasi-tetraploid cells (the number of chromosomes in the cells is twice that of normal animal somatic cells), their genetic properties are unstable. With each cell division, individual or part of chromosomes may be lost until the cells show a stable state.

When a sufficient number of stable monoclonal hybridoma cells are obtained, they are injected into the abdominal cavity of mammalian mice and allowed to grow to a certain extent, and these cells can be collected for massive production of monoclonal antibodies. Of course, this method of production is relatively primitive, and more modern production methods have emerged in recent years. The monoclonal hybridoma cells are transferred to a culture flask or a large cell culture tank for expanding culture, and then the cell culture medium is collected for production of monoclonal antibodies in large quantities.

2. Various Monoclonal Antibodies, Wide Applications

Although the emergence of monoclonal antibodies has only happened in the short period of several decades, the rapid development of monoclonal antibodies has brought great changes to the pharmaceutical industry.

According to the incomplete statistics, only monoclonal antibodies used in immunoassays in the United States have accounted for 30% of diagnostic testing items, and the profit reached 10 billion US dollars in 2000.

In 2010, the global therapeutic monoclonal antibody drugs had a sales volume of 44 billion US dollars, plus 10 million US dollars considering the monoclonal antibody diagnostic and research reagents, and the total amount of the monoclonal antibody drug market reached 55 billion US dollars. The total market for monoclonal antibodies in 2015 reached about 1 trillion US dollars. In the future, global monoclonal antibody drugs will still maintain a high growth rate.

Some scientists say that monoclonal antibodies are the first commercialized high-tech drugs, and the use of monoclonal antibodies instead of traditional clinical diagnostic methods has greatly accelerated the speed of clinical diagnosis. Through the specific reaction between the pathogenic antigens (such as pathogenic viruses, bacteria, and antigens on the

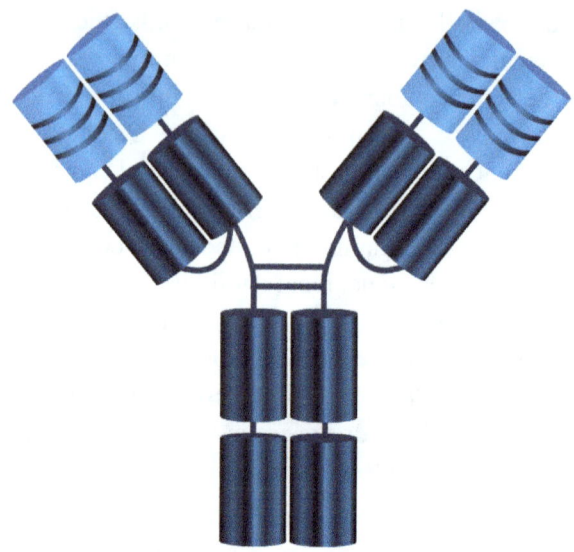

Model of monoclonal antibody molecule

surfaces of cancer cells) and the antibodies, it can be quickly diagnosed whether a person has a certain disease.

The antiserum used in the past was a mixture of various antibodies, and can only be used as an auxiliary method for diagnosis. To diagnose a certain disease, complex tests are still needed, which are time consuming and laborious, and often delay the treatment due to untimely diagnosis, and even cause death in serious cases.

Since there are monoclonal antibody diagnostic reagents for a single antigen, many diseases can be diagnosed in a very short time. It can be used not only in the diagnosis of viral diseases, bacterial diseases, sexually transmitted diseases, parasitic diseases, tumors, and other diseases but also in the diagnosis of immunodeficiency, early pregnancy detection, endocrine diseases of women, choriocarcinoma and hydatidiform moles, and cardiovascular diseases.

In recent years, the use of monoclonal antibody technology to produce highly sensitive diagnostic reagents has become more and more mature, and the products have been continuously simplified, and portable monoclonal antibody diagnostic kits have emerged. There have been many kinds of the kits in the Anglo-American markets. China has also produced monoclonal antibody diagnostic kits for hepatitis B with high diagnostic sensitivity and they have already been sold on the markets. In addition, there are more than a dozen products such as monoclonal antibody diagnostic kits for colon cancer and monoclonal antibody diagnostic kits for pancreatic cancer, which are in trial or have been put into massive production.

By so far, a series of monoclonal antibody reproductive health diagnostic test strips have been put into the markets in Yunnan province, which indicates that China's monoclonal antibody technology is already mature and has begun to enter the practical application stage.

At Yunnan University in Western China, researchers led by Dr. Lan Ma have been researching and developing monoclonal antibody technology for 10 years. By using hollow fiber bioreactor technology, antibody purification technology, immunochromatography colloidal gold coloration technology, test strip film and paper screening and matching technology, etc., they have successfully broken through the bottleneck of many technologies, and have mastered the full set of technologies from the large-scale production of monoclonal antibodies to the independent development and production of monoclonal antibody rapid diagnostic test strips. Great progress has been made in the production and application of domestic monoclonal antibodies in China. Yunnan province has the ability to produce 20 grams of monoclonal antibody per year and 80,000 test strips per day.

Monoclonal antibodies also perform well in helping organ transplantation. According to statistics, the annual incidence rate of renal failure in China is about 50–100 per million people. According to Leishi Li, an academician of the Chinese Academy of Sciences, the traditional treatment of renal failure is dialysis, but dialysis patients still have problems such as high cardiovascular complications, poor adaptability, and heavy economic burden. He said that allogeneic kidney transplantation is the most effective way to treat advanced renal failure, but it will cause immune rejection.

The humanized monoclonal antibody "Daclizumab" developed by the Roche company in Switzerland brings new hope to patients with acute rejection after renal transplantation.

According to the scientists, Daclizumab is the world's first monoclonal antibody for preventing acute rejection in patients with renal failure after transplantation. It is also the world's first humanized monoclonal antibody that specifically acts on the interleukin-2 (a lymphokine secreted by cells, which can regulate the function of immune system) receptor, and at the same time, it is also the first humanized monoclonal antibody approved by the US Food and Drug Administration for clinical application .

The mechanism of action of Daclizumab is to inhibit the activation and proliferation of lymphocytes mediated by interleukin-2 by blocking the binding of interleukin-2 and its receptor, thus inhibiting the cellular immune responses in transplanted organ immune rejection. International clinical data show that the incidence rate of acute rejection can be further reduced by 40%, the risk of one-year death of patients is reduced by 70%, and the risk of loss function after one year of kidney transplantation is reduced by 36%. And, there is no increase in side effects such as infection and lymphoma incidence rate. Chinese clinical trials have shown that the incidence rate of acute rejection is only 2.6% after the standard triple immunosuppressive regimen is added with Daclizumab, and there is no increase in serious adverse reactions such as infection.

In addition to detection and diagnosis, monoclonal antibodies can also be used as drugs for the treatment of diseases, mainly for cancer treatment, such as colorectal cancer, lymphoma, breast cancer, ovarian cancer, lung cancer, melanoma, leukemia, prostate cancer, and pancreatic cancer. There are also monoclonal antibodies for the treatment of rheumatoid arthritis and type 1 diabetes. Monoclonal antibodies can also be chemically coupled with various toxins (such as diphtheria exotoxin, ricin), radioactive elements, or drugs (such as aminopterin, doxorubicin) to prepare targeted drugs for the treatment of tumors, improve the efficacy of drugs on tumors, and reduce the side effects of drugs. In 2013, among the top 10 best-selling biologics in the world, antibody drugs accounted for five seats.

The growth trend of sales in the antibody drug market has not diminished, and countries around the world have invested heavily in it. The global pharmaceutical giants (such as Roche, Novartis, Pfizer) have spared no expense to develop antibody drugs. By the end of 2014, the China State Food and Drug Administration had approved a total of nine independently or jointly developed monoclonal antibody drugs, mainly in the fields of anti-tumor, anti-rejection, and autoimmune diseases, and some of them are still in the clinical research stage.

Monoclonal antibodies are not only used in medicine but also in agriculture. In recent years, there have been dozens of monoclonal antibody agents for the diagnosis and treatment of livestock diseases, including infectious anemia of horses, leukemia of cattle, foot and mouth disease, swine fever, and swine asthma.

It may not be unusual to treat animals, but even more interesting is the fact that monoclonal antibodies can also be used to diagnose and treat diseases in plants. Maybe you have not heard of it before. In fact, the principle is very simple; because plants and animals are similar, many diseases in both are caused by pathogenic microorganisms, such as bacteria, viruses, and molds. Monoclonal antibodies can accurately diagnose the strains of viruses and physiological races of bacteria, and can also be made into monoclonal antibody agents against a certain pathogenic virus or bacterium. This allows for rapid identification by taking pathogenic microorganisms from the lesion to find out the disease cause. There are many monoclonal antibody diagnostic reagents for crop diseases in the market, such as the potato virus (X, Y), tobacco mosaic virus, apple mosaic virus, citrus ulcer virus, and bacterial wilt.

So far, China has developed monoclonal antibodies against the potato virus (X, Y), equine infectious anemia virus, chicken Newcastle disease virus, swine blister disease virus, and foot-and-mouth disease virus type O. Among them, the monoclonal antibody diagnostic kits of potato virus Y and equine infectious anemia virus have been applied in practice.

Monoclonal antibodies can also be used in many fields such as basic research, industry, environmental protection and food testing, protein purification, and biological anti-cancer missiles.

3. Biological Missiles Eliminate Tumors

When talking about missiles, people will naturally think of the military missiles with steel bodies and huge lethality. It usually has nothing to do with a tumor. In fact, the biological missiles mentioned here are not exactly the same as the military missiles. The military missiles are developed by using rockets, which are a kind of weapon carrier with the function of navigation. They have strong recognition ability and can accurately hit the targets selected in advance. What are biological missiles?

As we know, the traditional treatment methods of tumors mainly include surgery, radiotherapy, and chemotherapy, but most of these methods cannot distinguish normal cells from tumor cells, and thus they cannot selectively destroy tumor cells. Some chemical drugs or radiation therapies can effectively eliminate a large number of tumor cells, but often also eliminate some normal cells, and in extreme cases may kill the patients. Even moderate doses of therapies sometimes may have strong toxic and side effects. Modern immunological studies have found that some monoclonal antibodies can recognize antigens on the surface of tumor cells and bind them firmly. In this respect, biological missiles are similar to military missiles. By using this characteristic of monoclonal antibodies, scientists can combine them with anti-tumor drugs to transport highly toxic anti-tumor drugs to tumor sites, killing tumor cells without harming normal cells. These conjugates of monoclonal antibodies and anti-tumor drugs are called biological missiles.

Once the biological missiles enter the human body, they can capture their own attack targets from many targets, and then follow the established route to destroy the enemy's bases — cancer lesions. Although military missiles have certain lethality, they often cannot completely destroy targets because of their insufficient power. They need to bring warheads with strong lethality, such as chemical warheads and nuclear warheads. The same is true for biological missiles.

Cancer cells proliferate rapidly; it is difficult to kill them when normal anti-cancer drugs are combined with monoclonal antibodies. In order to kill tumor cells, large doses must be used. However, on the one hand, this will inevitably cause toxic and side effects and pose a threat to healthy cells. On the other hand, scientists have found that the number of antigens on the surface of tumor cells is really limited, so the number of monoclonal antibodies that can be combined with them is not large, which also hinders the use of large doses of anti-cancer drugs. Therefore, when using biological missiles to treat tumors, it is important to choose the appropriate warhead. In practice, scientists have thought of using some highly toxic natural substances to replace anti-cancer drugs to make these warheads, such as bacterial toxins, ricin, and so on. Their combinations with monoclonal antibodies are also called immunotoxins because a very small number of immune molecules, even as little as one molecule, can kill a tumor cell. After the biological missiles are released, they can completely destroy the base of the cancer cells.

Although in theory, monoclonal antibodies can completely be used for cancer treatment, there are certain difficulties in clinical application, so the use of biological missiles to treat tumors or cancers has developed slowly in recent years. Fortunately, there have been some successful examples. In 1986, the world's first monoclonal antibody drug, the anti-CD3 monoclonal antibody OKT3, was approved by the US Food and Drug Administration, which was used to treat rejection after kidney transplantation. This was the prelude to the treatment of diseases with monoclonal antibody drugs. In 1990, Dr. Denador from the Davis University Hospital in California, USA, reduced the tumor size of six patients with advanced breast cancer that had spread by 50–75% using the monoclonal antibody method. The cancer in these six patients had invaded the chest wall or metastasized to the bone or lymph nodes. The monoclonal antibodies obtained from the experiments can fight against various cancer cells. It carries chemicals that can specifically kill targeted cancer cells and then be excreted outside the body without harming healthy tissues. Dr. Denador used radioactive iodine in combination with monoclonal antibodies to kill tumor cells. This drug, through metabolism in the body, can specifically target metastatic tumor cells at the bone, chest wall, liver, lymph nodes, and abdomen. In the study, the scientists found that if the patient received the lowest dose, the tumor could be reduced by 50%; if the highest dose was received, the tumor could be reduced by 75%. In 2002, the US Food and Drug Administration approved the first human monoclonal antibody, "Adalimumab Injection". This is a recombinant human tumor necrosis factor alpha monoclonal antibody expressed in Chinese hamster ovary cells for the treatment of rheumatoid arthritis and ankylosing spondylitis.

As of 2013, there were 46 monoclonal antibody drugs that had been approved by the US Food and Drug Administration, and more than 220 monoclonal antibodies entered the clinical trial stage. The therapeutic scope covers tumors, autoimmune diseases, organ transplant rejection, anti-infection, hemostasis, respiratory diseases, etc., among which the market for tumors and autoimmune diseases is the largest and the most diverse. In China, as of the end of 2014, the State Food and Drug Administration approved anti-human T-lymphocyte CD3 murine monoclonal antibody for injection, EnBoKe (Mouse Monoclonal Antibody Against Human Interleukin-8 Cream), YiSaiPu (Recombinant Human Tumor Necrosis Factor-α Receptor IgG Fc Fusion Protein for Injection), Vivatuxin (Tumor Necrosis Therapy Monoclonal Antibody Injection), LiKaTing (Iodine[131I] Metuximab Injection), Qiangke (Recombinant Human TNF Receptor-Ig Fusion Protein for Injection), TaiXinSheng (Nimotuzumab Injection), JianNiPai (Recombinant Humanized Anti-CD25 Monoclonal Antibody Injection, also called), LangMu (Conbercept Ophthalmic Injection), and other antibody drugs, of which Vivatuxin is used to treat liver cancer, LiKaTing is used to treat primary liver cancer, and TaiXinSheng is used to treat nasopharyngeal cancer.

A doctor at the Johns Hopkins School of Medicine in the United States combined radioactive iodine with monoclonal antibodies and injected them into patients with advanced lung cancer and achieved good results. Therefore, monoclonal antibodies as biological missiles are very promising to become powerful weapons against stubborn diseases such as cancers.

Of course, biological missiles have other uses. If other chemicals, such as biochemical reagents and reactants, are installed on warheads, they can not only be used to diagnose diseases caused by serum components, bacteria, viruses, parasites, etc., but can also be used to detect, isolate, and purify interferons, membrane proteins, and various trace components in cells that are difficult to purify by other conventional methods, so as to serve scientific research and medical treatment.

4. Potato-Tomato, New Species of Plants

A few decades ago, there was a period in which a number of academic journals in China translated and published news quoted from a foreign science journal describing the use of the cell fusion method by former federal German scientists to successfully cultivate an animal cell and plant cell hybrid called "beef-tomato". According to the news, the hybrid fruit of beef-tomato was disc shaped. This kind of tomato has both beef flavor and tomato flavor, and its nutrition is more comprehensive, because it contains animal proteins in its flesh, with animal proteins and plant proteins each accounting for half. What a wonderful fruit!

Potato-tomato is a super crop not found in nature

This science news was first published in a well-known foreign magazine on March 31, 1983.

As we all know, April 1st of each year is traditional April Fool's Day. On this day, according to the custom, people can lie and joke freely without any responsibility. So, the authenticity of the news was suspected at that time.

For the time being, whether this news is true or not, the scientific principle that contained in it is quite novel. The cells of different species can hybridize with each other to produce new species that have never been found in nature.

Although this principle is not realized between animal cells and plant cells, it has long been achieved between different plant cells. In our daily life, tomatoes and potatoes are often placed on our dinner tables. As we know, tomatoes and potatoes have very different shapes from each other. One is the tomato that blooms above the ground and bears similar round-shaped fruits; the other is the potato that grows under the ground and bears lumpy-shaped fruits. But, once upon a time, a miracle happened between them.

In 1978, German and Danish scientists fused the somatic cells of tomato and potato to form hybrid cells, and then managed to cultivate them into complete plants. As a result, a new hybrid plant "potato-tomato" with genetic characteristics of the two crops was born, which is unique in the natural world. It has both potato-like tubers and tomato-like branches. The branches bloom and bear bright red tomatoes, while the underground tubers are potatoes. The new variety, which is produced by cell fusion and hybridization, has both tastes. The appearance of potato-tomato, a two-story crop, has caused a sensation in the biological world.

According to scientists, in the nature world, it often takes tens of thousands of years to accumulate variations before one species can become another. The way that creates new species through cell fusion has greatly accelerated the evolution of species and enriched the treasure trove of species in nature. On the contrary, under natural conditions, it is almost impossible to form new species through sexual hybridization between two species with large genetic differences, which is called reproductive isolation in biology. It is produced in the long-term evolution of organisms, which helps to maintain the relative stability of species. Through cell fusion, this reproductive isolation can be broken and new species with relatively large variation can be created.

Unlike animal cells, plant cells are surrounded by a cell wall. Because of the existence of the cell wall, the two plant cells cannot fuse together. It is like chicken eggs. To fuse two chicken eggs together, you have to break the shells first. The plant cells with the cell walls removed are also called protoplasts. The earliest protoplasts were obtained by the mechanical method, but the efficiency was very low.

In 1960, Professor Cocking of the University of Nottingham, UK, first obtained protoplasts by the enzymatic method, which greatly improved the preparation efficiency. In 1971, Takebe and other scholars reported for the first time that a complete plant regenerated from mesophyll proto-plasts of tobacco was obtained. All these laid a solid foundation for the realization of the crossbreeding between distant higher plants. Up to now, more than 100 plants have been successfully cultivated by protoplasts in the world.

In the late 1960s and early 1970s, plant somatic cell hybridization tech-nology began to emerge. At that time, scientists saw the possibility of plant somatic cell fusion from the accidental fusion of Haplopappns gra-cilis and tobacco protoplast cells reported by Cocking and others. They gradually became interested in the research of this technology, hoping that it could break the barriers of distant hybridization. However, the experi-mental study of cell fusion technology started from two tobacco plants which could obtain hybrid offspring through sexual hybridization.

In 1972, American scientists fused mesophyll protoplasts of Nicotiana glauca and Nicotiana Langsdorffii to obtain interspecific somatic hybrid plants, which were almost the same shape as the offspring produced by sexual hybridization. After repeated experiments, it was proved that the experimental results were not a problem, thus causing a vibration in the biological world. Since then, many scientists have transferred their ener-gies to crossbreeding between distant plants.

The general process of cultivating hybrid plants by plant somatic cell fusion is as follows: first, two parent plants are selected, then protoplasts are isolated from their cells, then the two kinds of protoplasts are fused under the action of the fusion inducer, then hybrid cells are screened and cultured to regenerate new cell wall, and eventually the callus is produced through continuous cell division and differentiation to form somatic hybrid plants. Of course, in cell fusion for the purpose of breeding, the obtained hybrid plants should be observed and screened repeatedly.

Nowadays, more and more new plant species have been obtained by somatic cell hybridization technology. After the successful cultivation of potato-tomato by German and Danish scientists in 1978, scientists from the American Vanguard Corporation in 1982 also developed a hybrid of potato-tomato. It looks like a potato plant, but it has the excellent quality of resistance to fusarium wilt like a tomato plant. Later, California scien-tists also used the cell fusion method to produce tobacco resistant to triclosan. In 1986, Japanese scientists used the cells of purple cabbage and Chinese cabbage to form a new type of vegetable "Bio-White-Blue", which is similar to Chinese cabbage in shape and purple cabbage in taste. It belongs to interspecific somatic hybridization variety, and has advantages such as a short growth period, strong heat resistance, and storage suitability, which are widely welcomed by people. Before that, Japanese scientists successfully developed a valuable somatic hybridization variety between cultivated tomato and wild tomato, and obtained fruits. The scientists from the Department of Agronomy of Hokkaido University in Japan have transferred the protein genes of soybean to rice seeds to cultivate "soybean-rice". The soybean-rice has both the nutrition of soybeans and the function of rice to satisfy the hunger. This newly developed nutritious rice variety is richer in nutrients than the rice currently consumed. Brazilian agricultural expert Luis Mareto has transferred the common soybean protein genes into the Brazil nut and bred a Brazil nut soybean new variety by hybridization. It is rich in methionine and can be propagated rapidly. In 2011, Li Jiang and others reported the preparation, fusion, and regeneration of protoplasts of Agrocybe aegerita and Coprinus comatus. In 2012, Ji'an Wang and others reported that through the asymmetry fusion technique of protoplasts, it can not only maintain the favorable genes of soybeans but also introduce some favorable traits of other crops and plants, so as to cultivate excellent new soybean varieties. In 2013, Jihua Su and others reported the use of polyethylene glycol to induce protoplast fusion of wheat and aegilops, and established highly efficient technical systems for the heterokaryon formation to provide intermediate materials for wheat disease resistance breeding. In 2015, Jiejie Chu and others reported that high-efficiency fermentation of the *Saccharomyces cerevisiae* strain was selected using the protoplast fusion technology of inactivated biparental strains. The fermentation rate was increased by 108% compared with the fusion parents, and the diacetyl content was reduced by 60.7%. In the same year, Di Wang and others reported that by using DHA-producing Schizochytrium B4D1 and Aspergillus niger CGMCC 3.316 as the starting strains, protoplast fusion technology was applied to select new Schizochytrium that can be fermented using starches to produce brain gold DHA (docosahexaenoic acid). In 2016, Xiaolun Hou and others reported that the production of glutamine transaminase of Streptomyces maoyuanensis was improved by protoplast fusion technology.

Other hybrid cells cultivated by somatic cell fusion technology include "tobacco-soybean", "horsebean-petunia", "sugarcane-sorghum", "carrot-celery", "common-tobacco-yellow-flower-tobacco", "mushroom-cabbage", "arabidopsis-rape", "tobacco-petunia", "tobacco-fairy", and "kelp-undaria". Among them, common-tobacco-yellow-flower-tobacco, tobacco-petunia, tobacco-fairy, and kelp-undaria were first cultivated by Chinese scientists.

Although there are still some problems with distant hybrid plants such as potato-tomato, such as they cannot be promoted in agriculture (just like the infertile mules produced after mating horses and donkeys, the pollen of potato-tomato is infertile, and when light is insufficient, the fruit of potato-tomato will be small), it is a new type of plant on earth, which is amazing for the world. From it, people have seen the enormous potential of human beings to create new life.

With the development of sciences, the large-scale reproduction of potato-tomatoes will eventually be solved. For example, the rapid production of tube seedlings by plant somatic cell cloning technology may be a way.

5. Tumor Vaccines, Defeating Cancers

Cancer is a well-known enemy of human health. Many people will be nervous when they talk about cancer. Traditionally, the treatment methods of cancer include surgery, chemotherapy, and radiotherapy. The emergence of tumor vaccines undoubtedly adds new weapons to the fight against cancers.

Biologists have found that in order to be recognized and killed by the body's immune system, tumor cells must provide not only antigens (proteins on the surface of cells) but also stimulus signals. However, the cunning tumor cells can escape the blockade and kill off the human immune system by reducing the number of surface antigen molecules. At present, few tumor antigens have clearly been identified, and it is very time-consuming and inefficient to produce antigens by single-gene transfection, which makes the development of tumor vaccines very difficult.

The immune system in the human body produces dendritic cells after being stimulated by antigens, and the immune function is performed through dendritic cells. Thus, the dendritic cells are specific immune sentinels in the human body. In the early 1990s, researchers invented a new technology for cultivating dendritic cells in the laboratory while observing the immune system attacking cancer cells. They put the immune cells in tumor antigens and created the first dendritic cell vaccines. Preliminary experiments have shown that patients with lymphoma, malignant melanoma, and prostate cancer show a strong anti-cancer immune response after inoculation of dendritic cells that have been previously treated with a known tumor antigen. This is very exciting.

In early 2000, Alexander Kugler and his colleagues from the University of Göttingen in Germany made major breakthroughs in tumor vaccine research. They used weak electrical impulses to fuse human tumor cells with immune cells to synthesize a special vaccine that can fight cancer. After injecting this tumor vaccine into 17 patients with renal cancer that had already spread, they were pleasantly surprised that seven patients had tumor immune response. Under normal circumstances, if these renal cancer patients had been treated by traditional methods, one could only guarantee a survival rate of 10%. This undoubtedly brings new hope to patients.

Kugler has also developed a melanoma fusion vaccine. Clinical trials have shown that approximately 40% of patients generated tumor immune response, which is similar to the response rate of the renal cancer vaccine.

Khuff, another scientist working on cancer vaccine research, believes that for cancer vaccine research, researchers should be most concerned about whether it will stimulate the body's immune system to attack healthy tissues. Fortunately, after receiving a fusion cancer vaccine, no autoimmune adverse reactions have been found so far, whether in animal experiments or human trials. In addition, the fusion of melanoma and renal cancer vaccines was prepared at the Chariot Hospital of Humboldt University in Berlin, Germany. Clinical trials confirmed that the anti-tumor effect was obvious, and no significant adverse reactions were observed. This shows that when used in cancer treatment, tumor vaccines are not only stable in efficacy but also safe for patients.

The research group led by Khuff had long hoped to develop a tumor vaccine with individual characteristics, which does not need to identify specific tumor cell antigens. In 1997, they fused dendritic cells with cancer cells. At the same time, it is theoretically speculated that the dendritic cells after fusion will enable the human body to respond to a variety of tumor antigens, including tumor cell antigens that have not yet been discovered and isolated. Khuff has begun the trials of a hybrid cell vaccine for breast cancer patients. "It's clear that autoimmunity is possible, but it hasn't happened yet. It's really good news that no one knows why," he said.

Khuff predicted that scientists will seek ways to make more cancer patients develop anti-cancer responses through vaccines. He said, "There are many research programs that can make this method more powerful."

After Khuff's animal experiments were successful, Kugler immediately led a research group to similarly vaccinate renal cancer patients because renal cancer cell antigens had not been isolated at that time. They fuse healthy human dendritic cells with tumor cells, and hope that this hybrid can arouse the enthusiasm of the immune system more than the pure one synthesized by the patient's own dendritic cells. At the same time, in order to prevent cell growth from being out of control and creating new tumor threats, scientists have irradiated the hybrid cells with radiation before they are safely inoculated into the human body. One of the great advantages of this method is that they are not vaccinated with a specific antigen, but with all the components of the entire antigens of a cancer cell. The research group led by Kugler planned to directly compare the treatment criteria of renal cancer vaccines with that of standard chemotherapy and stimulating immune chemicals such as interferon.

In China, tumor vaccines are the key developmental new drugs of biological products. At the same time, the development of tumor vaccine also ranks at an advanced level in the world.

Chapter 4

The Mixed Enthusiasm for Animal Cloning

1. Sheep Dolly, Animal Star

In the classic Chinese mythology *Journey to the West* written by Chengen Wu, a great writer of the Ming Dynasty in China, every time in an emergency, the monkey king Wukong Sun, who takes the golden cudgel and has the ability to make seventy-two changes, will pluck a vellus hair from his body, and then duplicate himself in large quantities by gently blowing on the vellus hair. Although this is a myth, with the development of sciences and technologies, it can be realized theoretically today.

Because the monkey's vellus hair is derived from the skin, it consists of cells, each of which contains all the genetic information which can develop into a complete monkey. This is because during the embryonic development process, in the cells that develop into vellus hairs, due to the influence of various factors on the activity of genes, only the genes related to vellus hair development synthesize proteins, while other genes are in a turned-off state. If these turned-off genes are activated, the vellus hair cells can develop into complete animal individuals like fertilized egg cells. The principle is simple. If a cell is taken from a plant leaf and properly cultivated, it can grow into a plant with identical heritability. However, for a long time in the past, it was widely believed in the biological world that once the animal cells differentiated (that is, develop into specific functional cells, grow into tissues or organs), the totipotency is gradually lost. Because of this traditional concept, most scientists have not dared to come into this "forbidden zone of science" of animal cloning.

The cloned sheep Dolly

In July 1996, for the first time in the world, a bleating lamb was cloned from the mammary cells of adult animals by Ian Wilmut, a well-known biologist at the Roslin Institute spinoff PPL Therapeutics in Edinburgh, the capital of Scotland and a famous cultural ancient city. After the birth of this little lamb, the scientist named it "Dolly" by the name of a very famous American singer, just like one's own child. The reason why it was so named was that scientists initially hoped it could become a famous animal star, but they did not expect it to live up to their expectations. Once this achievement of the cloning technology was published, it immediately caused a global stir. The ewe Dolly became the most dazzling star in animal history, which made the media all over the world crazy [11]. For a while, this little lamb with a mother but without a father was in the limelight.

One very important reason why Dolly caused a worldwide stir is the conceptual breakthrough. Because, before this, whether it was in the biological books of middle schools or universities, it was impossible to reverse the differentiated cells. Before that, even scientists who had been studying cloning technology for many years believed in this theory. Therefore, when cloning, everyone would choose to use embryonic cells as experimental materials. Dolly is the first cloned sheep by using a well-differentiated mammary cell taken from an adult animal as experimental material. This is a major breakthrough in theory.

The article about the success of sheep Dolly's cloning was published on February 27, 1997, and in the world-famous academic journal *Nature*, which is published in Britain. Its birth proves that the so-called highly differentiated cells, which perform specific functions and have specific shapes in animal bodies, have the potential to develop into complete individuals like fertilized eggs. In other words, animal cells, like plant cells, also have genetic totipotency, thus breaking the shackles of traditional ideas. This is very commendable, which also made this achievement rank first in the world's top ten scientific discoveries by American weekly journal of *Science* in 1997.

So, what is cloning? To put it simply, cloning is asexual propagation. For example, under suitable conditions, an *Escherichia coli* cell can be divided from one into two in 20–30 minutes; when a willow branch is cut into 10 segments, each segment may grow up into a willow tree; cut a potato into pieces and they all can take roots in the soil; the branches of the Scindapsus (*Epipremnum aureum*) are cut into several segments, and each cutting can also take root...These breeding methods are all based on the fact that the organism can divide into two or use a small part of its body to reproduce the offspring. It is completely different from the sexual propagation of higher organisms and is an asexual propagation method.

The English word "clone" originated from the Greek word "Klone", which means propagation by twigs or cuttings. According to the American National Bioethics Advisory Commission, the term "cloning" refers to the precise genetic replication of molecules, cells, plants, animals, or humans. However, the European Commission believes that "cloning" refers to a method of producing genetically identical organisms. "Cloning" is asexual propagation, but it means not only asexual propagation but also any group of individuals who come from a common ancestor through asexual propagation. The progeny group from the asexual propagation of a common ancestor, also called clonal propagation, is referred to as a clone. The genetic composition of all members in the same clone should be identical, with the exception of gene mutation. Clones of natural plants, animals, and microorganisms already exist in nature, for example, identical twins are actually a kind of clone. However, the occurrence rate of this kind of natural mammalian clone is very low, the number of members is too small (generally two), and due to a lack of purpose, it is rarely used for the benefit of human beings. For this reason, people began to explore the use of artificial methods to produce higher animal clones. In this way, the word "clone" began to be used as a verb, referring to the act of artificially breeding cloned animals.

In nature, some animals normally rely on the fusion of the sperm (male cell) produced by their father and the egg (female cell) produced by their mother to form a fertilized egg (zygote), then gradually develop into an embryo through a series of cell divisions from the fertilized egg, and finally form a new individual. Like this, breeding through the fertilization of the two kinds of sexual cells provided by parents to reproduce the next generation is called sexual propagation. However, if we use surgical methods to cut an embryo into two, four, eight, sixteen pieces to make an embryo grow into two, four, eight, sixteen organisms through special methods, to put it bluntly, these organisms are clonal individuals. These two, four, eight, sixteen individuals are called clonal lines (also called clones). In fact, a large group of monkeys, which are transformed from a monkey vellus hair in *Journey to the West*, is called cloned monkeys.

How was the cloned sheep Dolly born? Ian Wilmut and other scientists first injected hormone gonadotropin into the Scottish black-faced sheep to induce it to ovulate. Immediately after the egg was obtained, the scientists took out the nucleus from the egg cell with a very small pipette, while taking out the nucleus of the mammary cell of the Finn Dorset 6-year-old ewe, which had been pregnant for three months, and immediately sent the nucleus of the mammary cell into the egg cell. After the operation was completed, the egg cell was stimulated by electric pulses of the same frequency, so that the egg cytoplasm of the Scottish black-faced sheep and the nucleus of the mammary cell of the Finn Dorset were coordinated and adapted to each other. They let this man-made cell divide, develop, and eventually form an embryo in a test tube like a fertilized egg.

The embryo was then skillfully transferred into the womb of another ewe selected as the surrogate mother. After transplantation, the embryo developed well. In July 1996, the surrogate mother gave birth to the little sheep Dolly. Dolly is not the product of fertilization of the ewe's egg cell and the ram's sperm cell, but the result of gradual development of a nucleus-transferred egg, so it is a cloned sheep, which is different from the sheep born after mating with a normal ram and ewe.

The cloning process of Dolly the sheep seems simple, but the operation is much more complicated. Just for a successful nucleus transfer, it was repeated 277 times, accounting for 63.8% of the total number of the nucleus-transferred eggs. When cultivating these valuable nucleus-transferred eggs, only about one tenth (29) of them are viable and can grow into the morula stage or blastocyst stage of embryo development. When these 29 early-stage sheep embryos were transferred separately into the uteruses of 13 different sheep selected as the surrogate mothers, only one lamb "Dolly" was successfully born. It can be seen that the success rate is extremely low, indicating that the cloning experiment was very difficult at that time, and the cloning technology was not mature enough then.

The idea of cloning animals using nucleus transfer technology was originally put forward by Hans Spemann in 1938, which he called "bizarre experiment". That is to say, the nucleus is taken out from the embryo (either mature or immature embryo) in a later development stage, and transferred into an egg. This idea is also used in the cloning of Dolly the sheep.

Since 1952, scientists have used frogs to carry out nucleus transfer cloning experiments, and have obtained tadpoles and adult frogs. In 1963, the research group led by Professor Dizhou Tong, a famous Chinese scientist, first studied the cell nucleus transfer technology of fish embryos with experimental materials, such as goldfish, and achieved success. The first achievement of the study of mammalian embryonic cell nucleus transfer was obtained in 1981 by K. I. Illmense and P. C. Hoppe using mouse embryonic cells to produce normal-developing mice. In 1984, Danish scientist Stern Villadsen successfully cloned a sheep by using an immature embryonic cell. Later, other people used pigs, cattle, goats, rabbits, macaques, and other animals to repeatedly verify his experimental method. In 1989, Villadsen obtained the second generation of cloned cattle with the continuous nucleus transfer technique. In 1994, Neil obtained cloned cattle from the late-stage embryos with at least 120 cells. In 1995, embryonic cell nucleus transfer was successful in major mammals, including frozen and *in vitro* produced embryos, and nucleus transfer experiments on embryonic stem cells or adult stem cells were also tried. Unfortunately, till 1995, the nucleus transfer of differentiated adult animal cells had not been successful. Therefore, Dolly is a landmark precedent in the history of animal cloning, with unusual significance.

Many years later, Dolly, the most photographed star in the history of animals, became a mother. After the first birth of the female lamb, Bonnie, on April 13, 1999, two male lambs and one female lamb were born on March 24, 2000. After giving birth, Dolly and her children lived in a special fence with infrared heating equipment at the Roslin Institute Farm in Edinburgh, UK. They were all in good health and the lambs could also milk normally, which shows that Dolly was fully qualified to be a mother despite her special status. According to information, the three newborn children of Dolly share the same father as their sister "Bonnie", a common Welsh male goat named "Davi". After that, Dolly gave birth to three more lambs.

In February 2003, veterinarians found that Dolly had suffered from severe progressive lung disease, so she was euthanized by the investigator because it was an incurable disease. According to the Roslin Institute, Dolly had been coughing for a week before being diagnosed. Dolly's body was made into an animal specimen and stored at the National Museum of Scotland as a symbol of 20th century technology.

In September 2008, Professor Ian Wilmut, together with Dr. Campbell, the cocreator of Dolly, and Professor Shinmi Yamanaka of the Kyoto University in Japan, who invented induced pluripotent stem (iPS) cells, jointly won the "Oriental Nobel Prize", which is known as the Yifu Shao Life Science and Medicine Award. Regrettably, Ian Wilmut missed the 2012 Nobel Prize, which surprised many people.

The birth of the cloned sheep Dolly not only proves the genetic totipotency of animal somatic cells but also opens a new chapter in cloning animals especially mammals by human beings, which is a major event in the history of biology.

2. Xiangzhong Yang and Cloned Cattle

The cloned cow Amy was born a few years after the cloned sheep Dolly had appeared. Its owner is Professor Xiangzhong Yang, a famous Chinese-American scientist. In 2001, he was received by Jiabao Wen for unlimited time, then Vice Premier of the State Council of China. Professor Xiangzhong Yang, who was born in Wei County, Hebei Province, China, once worked as a barefoot veterinarian in his hometown, and was then admitted to the Department of Animal Husbandry and Veterinary Medicine of Beijing Agricultural University, and went to the United States after graduation. Professor Xiangzhong Yang has been doing animal cloning research since 1983 and worked at the Animal Transgenic Center of Cornell University. But, in the early stages, he cloned animals by embryo technology. After 1997, he turned to cloning using adult animal somatic cells. The achievements that he made were very remarkable, and he was known as the father of the cloned cattle throughout the world.

According to media reports, Amy, a cloned cow, was born in June 1999, which was quite dramatic. Professor Xiangzhong Yang said that after many cloning failures, several cows finally became pregnant. On June 10, a pregnant cow was about to give birth, about three weeks ahead of its due date. At that time, Professor Xiangzhong Yang's postdoctoral researchers and veterinarians called him from the farm to tell him that the pregnant cow was in labor. Professor Xiangzhong Yang didn't expect this, but he was still very excited when he thought about the birth of cloned cattle.

At that time, Professor Xiangzhong Yang was driving with his wife to primary school to see their child, but soon after, the farm called again and told him that it might be bad. The veterinarian found that there was no fetal sound, and everyone was very nervous. Professor Xiangzhong Yang was so nervous that he couldn't even drive the car. He thought that a cow was going to die. After that, Professor Xiangzhong Yang returned to his office in a daze and sat in his chair. He was very sad to think that the cow was dead.

When he was feeling sad, the farm called again. This time, it was a veterinarian. Unexpectedly, it was good news. The veterinarian said that they had a caesarean section. After opening the cow's stomachs, they found that the calf was not only alive but also very well. Everyone was very excited. After answering the phone, Professor Xiangzhong Yang couldn't wait. He went to the scene immediately. The cow was already on the ground.

Professor Xiangzhong Yang was so excited at that time!

Amy was a heifer. She was born at 94 pounds. She was as healthy as a normal cow. From the pattern, it was basically the same as the mother. It was a black and white cow, especially beautiful. Seeing this, Professor Xiangzhong Yang was too excited to speak.

Talking about the naming of the newly born cloned cattle, Professor Xiangzhong Yang said that the name "Amy" had no special significance. At that time, he mainly considered that there were still more than a dozen pregnant cows and wanted to rank them from A, B, C, and D. As a result, some people joked with him, saying that the name of the cloned cattle could be better, and if it was better, it would be more famous.

Professor Xiangzhong Yang asked, "How can we be better?" The Joker said, "Look at the cloned pigs. There are five pigs in total. They can't remember any other names. The only name you can remember is ".com". Because it is a popular internet domain name, everyone remembers it. And Dolly, why do you remember it? Because it is the name of a very famous American singer.

Compared with the cloned sheep Dolly, the cloned cow Amy made a great breakthrough in technology. Before Amy appeared, scientists used epithelial cells of the mammary gland, granulosa cells of the ovary or oviduct cells to clone, which were all related to germ cells. Therefore, it was natural for people to think that the success of cloning was probably due to the fact that germ cells are connected with the reproductive system. Many foreign scientists and even academicians thought so. However, Amy was cloned from bovine ear skin cells that had nothing to do with the reproductive system. This is actually a conceptual breakthrough, just as Dolly, which was the first to use adult animal somatic cell to clone. Therefore, in terms of technology, the cloned cow Amy is similar to the cloned sheep Dolly, but there are still great improvements. This improvement greatly improves the efficiency of cloning. The success rate of the cloned sheep Dolly was 0.36%, while that of the cloned cow Amy was 10%. The cloning efficiency was increased dozens of times, which is a kind of technical progress.

Professor Xiangzhong Yang also said that their inspiration for successfully cloning Amy this time came from their failure and from the process of having suffered many losses. After the cloned sheep Dolly appeared, he thought that many scientists would be like him, regretting that he did not become the first sheep cloning or other animal cloning success in the world because his own laboratory and other laboratories of the same profession have done similar research as Dolly the sheep, using the cell nuclei of adult animals for cloning. Although he got embryonic development, it was a pity. Because he did not believe that the differentiated somatic cells might be reversed, he did not go to transplant. If he had carried out further transplantation, he might have been the first person to perform sheep cloning in the world, and he would have had a sensational effect like Dolly. The reason why he suffered losses was because he believed too much in books. He learned a lesson when he cloned Amy the cow. Even if others say that only germ cells and embryonic cells can be used, he still doesn't believe it. Instead, he thinks that the cells of adult animals also contain all the genetic information of life, that is, any cell in an animal should contain the same genetic information. On April 16, 2002, the cloned cow Amy finally gave birth to a healthy male calf weighing about 47 kilograms, named "Finally".

In fact, biological cloning is not a new topic. It has been done for nearly a century since 1905. During this period, it developed tortuously and slowly. Finally, a few years ago, British Ian Wilmut and others cloned Dolly the sheep with somatic cell, which attracted worldwide attention. In fact, the cloned cow Amy is not Professor Xiangzhong Yang's first cloned work. Before that, he had successfully cloned many calves in cooperation with his Japanese counterparts.

According to reports, Kamitakafuku is a very famous bull in Japan. As early as December 1997, Professor Xiangzhong Yang and his Japanese doctoral student Kubota Chikara took some ear skin cells of Kamitakafuku and cultured them *in vitro* for several generations, and then carried out the experiment of nucleus transfer.

At that time, Kamitakafuku was a 17-year-old cow (equivalent to 80–90-year-old human).

In December 1998, they finally cloned four calves from the ear skin cells of Kamitakafuku for two months, and two of them survived. In

February 1999, they cloned two calves from the cells cultured for three months, and both survived healthily. The four surviving bulls were named as Ichiro, Jiro, Saburou, and Shirou. Their English names are Tommy, Andy, Timothy, and Anthony (from TATA, which is the DNA transcription initiation signal). According to the date of birth, these cloned bulls may be the first male animals to be cloned successfully in the world.

The paper jointly written by Professor Xiangzhong Yang and others was published in the *Proceedings of the American Academy of Sciences* on January 4, 2000. Their research results are considered to be a major breakthrough in biological cloning research: first, this achievement breaks the traditional concept that only fresh cells or cells after short-term culture can be used for cloning, proving that long-term culture of somatic cells will increase the number of clones; second, till now, people had only used mouse stem cells to clone for research such as gene functions and changing animal genetic traits. The Enlightenment of Professor Xiangzhong Yang's research results is that people can consciously change the DNA of donor cells and clone animals with special characteristics by using long-term cultured cells *in vitro* to clone other animals. Third, the cloned cattle are old, and the cells taken from its ears can still clone offspring after long-term multi-generation culture, and both the donor and the cloned offspring are still alive, which provides a direct comparison possibility for studying the mechanism of aging and longevity. Fourth, it is of great significance for agricultural production and saving endangered wild animals, and it is also of great significance for the use of cloning technology as new approaches that have been developed in functional gene localization, diagnosis and treatment of genetic diseases, tissue and organ repair, and so on.

Professor Xiangzhong Yang's laboratory once produced a large number of cloned animals. In the experiments on cows, a total of 10 animals were born. In cooperation with his Japanese students, he produced dozens more. In total, there were more than 30 cloned animals, all of which were from old animals. As for why biological cloning always chooses elderly animals, Professor Xiangzhong Yang explained that this is because many diseases requiring organ transplantation, such as Alzheimer's disease and Parkinson's disease, mainly occur in the elderly. Cloning technology can be used to cultivate tissues and organs for transplantation therapy.

Professor Xiangzhong Yang also used cloning technology to study the function of genes. The traditional concept holds that once the cells have completed differentiation, the function of genes will be irreversible, and now this can be reversed after cloning. A skin cell can be transformed into an embryo and then become an individual. There is a process of cell dedifferentiation, and the process of turning on the genes that have been turned off.

Professor Xiangzhong Yang applied for patents for his cloned calves. He believes that the establishment of patented technology is conducive to stimulating the development of science and technology.

3. More and More Cloning

Since the birth of the cloned sheep Dolly, biological cloning has been very popular, resulting in more and more cloned animals, both in kind and quantity. In March 1997, just a few months after the birth of Dolly, scientists from the United States, China, and Australia published news of the success of cloning monkeys, pigs, and cattle. However, they were all cloned by using embryonic cells, and the significance is naturally not comparable to the cloned sheep Dolly. On July 9 of the same year, scientists at the Roslin Institute in Edinburgh successfully bred the cloned sheep Polly, which once again caused a global stir. Unlike the cloned sheep Dolly, the cells of the cloned sheep Polly carry a human gene that can produce an expensive protein drug.

According to the authoritative academic journal *Science* published in the United States, the cloning process of Polly sheep is as follows:

First, a part of the fibroblasts of the embryo is taken out and exposed to DNA solution containing the human genes and marker genes, so that the exogenous genes can enter into the embryonic cells. Then, culture and test are carried out to see which embryo cells have exogenous genes. And then, the cell nucleus containing the exogenous gene was transferred into the enucleated egg cell. The cell is fused by electric pulse to develop into an embryo, and eventually transferred to the uterus of a sheep as a surrogate mother. In this way, the born lambs may have exogenous genes.

In July 1998, Wakayama and others from the University of Hawaii reported that 27 surviving mice were successfully cloned from mouse cumulus cells, and seven of them were offspring cloned by the cloned mice. In addition, they used a new, relatively simple, and higher success rate cloning technique, which was named "Honolulu technology" after the location of the university. In January 1999, a mouse cloned with embryonic stem cell was born.

Since then, scientists in the United States, France, the Netherlands, and South Korea have successively cloned cattle using somatic cells. By the end of 1999, the cloned progenies of six types of somatic cells including fetal fibroblasts, breast cells, cumulus cells, fallopian tube (uterine epithelial) cells, muscle cells, and ear skin cells had been successfully born in the world.

The cell nucleus transfer experiments carried out between different species also achieved some gratifying results. In January 1998, by using cattle eggs as receptors, American scientists had successfully cloned embryos of five mammals, including pig, cattle, sheep, mouse, and macaque. The research results of this study indicate that unfertilized eggs of a certain species could be combined with mature nuclei from a variety of animals. Although these embryos were aborted, it made a beneficial attempt toward the possibility of heterologous cloning. In 1999, American scientists cloned embryos of rare animal argali with bovine eggs; Chinese scientists also cloned early-stage embryos of the endangered animal giant pandas with rabbit eggs, and these achievements may become a new way to protect and save endangered wild animals.

The cloned pigs used for organ transplantation have also come to the world. On March 14, 2000, the British PPL Biotech Co. announced that it had successfully cloned five piglets. Different from the previous cloning, scientists had modified the pig's genes before cloning, so the organs of these cloned pigs can be used for human transplantation, which is of great medical significance. Because pigs are prolific animals, they must have several viable embryos in their bodies to maintain normal pregnancy. But, sheep and cattle are different; they just need one embryo to maintain normal pregnancy, therefore, it is more difficult to clone pigs than sheep and cattle. Originally, only four embryos were found during the ultrasound scan, but five piglets were born, which is undoubtedly an unexpected pleasant surprise. Subsequently, the biotechnology company of PPL Therapeutics conducted clinical medical trials on cloned pigs.

In October 2000, a cloned pig called "Senna" was born in Japan. Japanese researchers used small needle-like tubules to transfer as many as 100 black pig embryos after special treatment into four sows by microinjection. Finally, only Senna survived. Microinjection technology can precisely select which genetic materials to transfer, and even separate certain chromosomes, and avoid contamination of the laboratory supplies with the cell nucleus or genetic material of the cloned object.

In December 2000, the well-known Roslin Institute in Edinburgh cooperated with the American Viragen biotechnology company. After more than two years, special chickens with genetic modification were cloned and one of them was named "Britney". This cloned chicken can be used to produce anticancer drugs. According to the British *The Mail on Sunday*, the protein content of hens is determined by the genes of hens. The eggs laid by hens will contain a large number of specific proteins by changing the genes in one cell nucleus. It was by modifying the genetic material in a single cell nucleus that the researchers created the genetically modified hens. The eggs laid by hens contain a lot of proteins needed by researchers. Each cloned chicken can lay 250 eggs a year. Each egg contains at least 100 milligrams of proteins that are easy to extract and can be used to make anticancer drugs. These proteins could only be produced in the laboratory before, and even if only a small quantity, the production of them is very difficult and expensive. This hinders the development of new drugs for a variety of diseases, including breast and ovarian cancer. This new generation of anticancer drugs from cloned chicken has broad clinical application prospects.

At the same time, bovine cloning was also moving toward a practical direction. In July 2000, Xinhua News Agency reported that a female somatic cell cloned cow gave birth to a calf in the Japanese Ishikawa County Animal Husbandry Comprehensive Center, which proved once again that somatic cell cloned cattle have normal fertility. According to the information provided by the Department of Agriculture, Forestry and Fisheries of Ishikawa County, this calf is female. The somatic cloned cow "Kaga 2" from the Ishikawa County Animal Husbandry Comprehensive Center was born naturally. Its body weight was 26.5 kilograms and the body length was 53 centimeters. It could stand up only 10 minutes after birth and then went to breastfeed. Mother and son were safe and in good health.

The Ishikawa County Animal Husbandry Comprehensive Center is the world's first agricultural research institutions to clone cattle using somatic cell cloning technology. "Kaga 2" is the second somatic cell cloned cow of the center, born on August 8, 1998. On September 30, 1999, researchers at the center used frozen semen from common breeding cattle (black-haired Japanese cattle) to make cows pregnant by artificial insemination. A cute calf was born on the expected day of delivery. The head of the center said that it was the first time in the world for a female cloned cow to give birth. This indicates that female somatic cell cloned cattle have normal fertility. In addition, in late January 2000, the Kagoshima County Institute of Food Cattle Improvement used the somatic cells (ear cells) of the cloned cattle to clone the next generation, which is so-called "re-cloned cattle". The results showed that the cloning technology of cattle was becoming more and more mature.

Primates also appear in the large family of cloned animals. The American journal of *Science*, published on January 14, 2000, reported that American scientists had successfully cloned a little monkey using asexual reproduction technology. Once this news was released, it immediately attracted great attention from all walks of life and the scientific community. According to CNN, a group of scientists began to study cloning technology a few years ago and planned to clone monkeys more than a year ago. As a result, after repeated failures, they finally succeeded. They named the cloned monkey "Tetra", a female short-tailed rhesus macaque. The cloned monkey Tetra was born at the Schatten Laboratory in the Oregon Regional Primate Research Center in Beaverton, USA. The scientists also provided the media with its photo at 4 months old. According to the reports, the laboratory adopted a new clonal method, named "embryo splitting". When the embryonic cell divides into 8 cells, it is divided into four parts, and then cultured separately, and finally four identical new individuals are produced. Since monkeys belong to the primates and are also the closest animals to humans, this research result means great progress in cloning technology, which actually indicates that there is no technical barrier to human cloning.

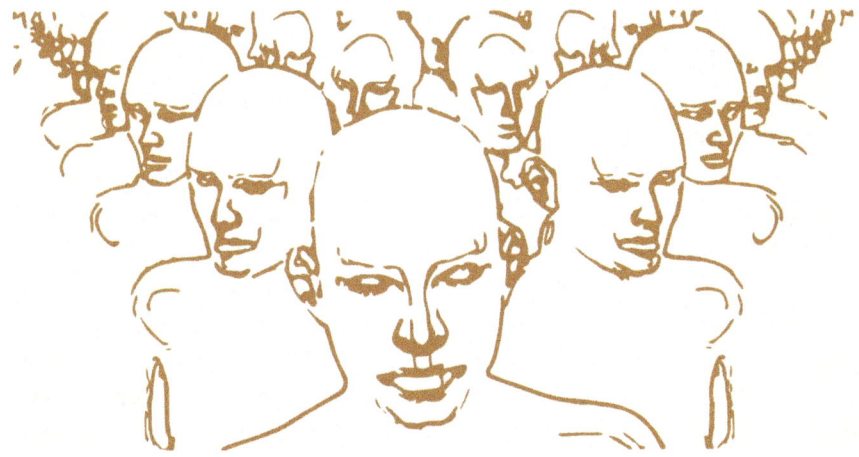

The cloned human beings

In the global enthusiasm of cloning, Chinese scientists are also unwilling to be left behind, and cloned a batch of animals.

On November 15, 2000, researchers from China and France carefully cultivated a somatic cell cloned cattle, using the nucleus-injected transfer technology invented by Chinese scientists. It is the world's first somatic cell cloned cattle that uses the ear cell of an adult cow, for which the adopted clonal techniques are different from the cloned sheep Dolly. The calf was female with the birth weight of 51 kilograms and all the health indicators were normal. Professor Qi Zhou, from the Institute of Developmental Biology of the Chinese Academy of Sciences, who participated in the study, said that the process of introducing exogenous genetic material can be simplified from two steps to one step by means of nucleus-injected transfer technology instead of the electrofusion method used in the cloned sheep Dolly. Not only can it shorten the production cycle, improve the efficiency of cloning, but also control the activation time of the cloned embryo and adjust its cell cycle. The previous attempts of non-electric fusion technology all ended in failure. As early as 1999, Chinese and French scientists successfully bred the world's first adult somatic cell cloned mouse using mouse embryo micromanipulation damage excision invented by Chinese scientists and the molecular biology techniques mastered by French researchers.

Cloned cattle further promoted the practical application of cloning technology. In the future, it will no longer be a fantasy to clone a large number of excellent breeding animals, which is helpful to promote the rapid development of animal husbandry.

On December 24, 1999, Hebei Agricultural University and the Biotechnology Research Center of Shandong Academy of Agricultural Sciences jointly tackled the key problems and successfully cloned rabbits in Jinan, China. Two cloned white rabbits were named "Luxing" and "Luyue", both of which grew well. In 1998, Dr. Wancun Chang, who graduated from Northwest Agricultural University, came to work in the Biotechnology Research Center of Shandong Academy of Agricultural Sciences. Under the guidance of well-known experts in the world, Dr. Wancun Chang worked with his assistants for many experiments. With the unremitting efforts of nearly one year, he successively overcame technical difficulties such as microinjection, electroactivation, electrofusion, *in vitro* culture, and *in vivo* transplantation. Finally, Luxing and Luyue were born. Although they were born from a gray rabbit, in fact, their real mother was a New Zealand white rabbit. Within 65–70 hours after mating with another male rabbit of the same race, the technicians extracted the embryo that was undergoing the fourth division from the uterus, and separated it into individual blastomeres by enzyme digestion and extracted the nucleus. At the same time, the researchers took out the oocytes of another New Zealand white rabbit, removed the nucleus of these cells, then processed them through microscopic injection, electrofusion, and other techniques, and then merged them into recombinant embryos and transplanted them into the big gray rabbit. A month later, the big gray rabbit gave birth smoothly to these two lovely little white rabbits.

This experiment belongs to embryo cloning, which has not yet reached the level of somatic cell cloned Sheep Dolly, but it has laid the foundation for the progress of cloning technology in China. At the same time, it provided an effective method for animal breeding, which can achieve the "factory" production of embryos and the optimal combination of genes.

On October 15, 1999, the first cloned goat of the somatic cell of transgenic goat was born in Yangzhou, Jiangsu Province, China. This was completed by the Institute of Developmental Biology of the Chinese Academy of Sciences in cooperation with Yangzhou University. The cloned goat was white and weighed 16.5 kilograms, and the main organs such as the heart, liver, and lung were normal. It was very active in the flock. Its scientific significance is extraordinary, because a drug gene had been introduced in the goat, making it a living animal pharmaceutical factory.

Scientists introduced the drug genes that people need into the fertilized eggs of animals. As the fertilized eggs divided, the introduced genes were multiplied along with the chromosomes in the cells, and could be stably inherited to the next generation. The animals carrying drug genes are called transgenic animals.

Transgenic technology can make mammals such as cattle and sheep become medicinal animals. However, in order to produce this kind of medicinal animal stably and massively, cloning is necessary. During the experiment, scientists also invented a new preparation technique. The biggest difference between this technique and the previous method of somatic cell cloning is that the latter must use transgenic goats to replicate offspring, and the probability of success is relatively low. However, this preparation technique uses ordinary goat cells and injects the needed genes into the cells. After confirming that the cells do carry the gene, the somatic cell nucleus can be transferred into other goats, so that the born goat would also be a transgenic goat. Its success rate even exceeds the technique of Dolly's cloning. This transgenic goat will be expanded into a group in the short term, and the latter can even expand indefinitely. Under the impetus of the local government, the Jiangsu Province Transgenic Animal Pharmaceutical Center was founded in Yangzhou University, and experts are promoting the development of the bio-pharmaceutical industry.

In Xi'an, a famous historical and cultural city, the work of cloning goats is also in full swing. The "Study on Somatic Cell Cloned Goat" is a key project of the National Natural Science Foundation of China and a key project of the Ministry of Agriculture of China. This program is presided over by academician Yong Zhang, who is the Professor, doctoral supervisor, and director of the Institute of Biotechnology of Northwest A&F University. On January 16, 2000, the researchers transplanted cell embryos and blastocysts grown *in vitro* into a Guanzhong dairy goat, and a few months later the goat gave birth to a cloned goat called "Yuanyuan".

On January 26, the researchers transplanted the cell embryo and blasto-cyst into a Saanen dairy goat at the Northwest A&F University, and the receptor goat gave birth to the cloned goat "Yangyang". Yuanyuan and Yangyang are both females, and the somatic cells used for cloning were taken from the same grey goat, so the sisters look exactly the same.

The cloned grey goat "Yuanyuan" and its surrogate mother

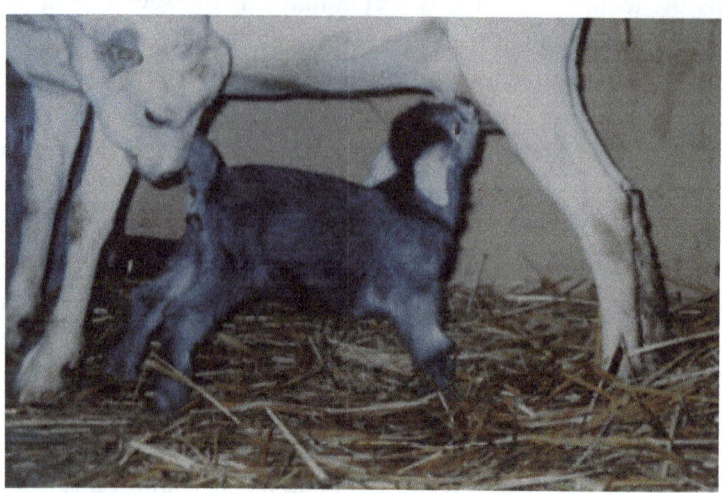

The cloned grey goat "Yangyang" and its surrogate mother

The 15-year-old cloned goat "Yangyang"

On June 22, 2015, the cloned goat Yangyang celebrated her 15th birthday. Yangyang, who has been in the same hall for five generations, has lost all her teeth. The average life span of a goat is 16–18 years. The 15-year-old Yangyang is considered to be the "old longevity star" in goats. It has visited Beijing three times, participated in the Beijing International Exposition, the National 863 High-Tech Exhibition, and has also done special programs on the China Central Television.

In December 2015, China Boyalife Stem Cell Group Co., Ltd. cooperated with Sooam Biotech Research Foundation of Korea to build the world's largest cloning factory in Tianjin, China. It plans to produce 1 million cloned cattle each year and realize the industrial production of cloned animals.

4. Cloned Animals Have Defects

As for cloned animals, people are generally concerned about whether the cloned animals with a mother but without a father have the same physiological functions as normal animals with parents? According to the Scottish scientist who has successfully bred Dolly, the world's first somatic cell cloned sheep, "Dolly the sheep showed obvious premature aging symptoms when she was only a few years old." Premature aging has become one of the defects of some cloned animals.

The average sheep can live for about 12 years, while Dolly, a cloned sheep, only lived for 6 years, which is only half the life span of ordinary sheep. Its early death has raised concerns about whether all cloned animals will prematurely age. How is the age of cloned animals calculated? Whether it is calculated from the age of zero or from the cumulative age of the cloned animal plus its parent animal, or from a certain age between the two, it is worth further study. As for the cloned sheep Dolly, whether she was born at the age of 0 or 6, or a certain age between the two, there is still no conclusion. On February 14, 2003, the cloned sheep Dolly died of pulmonary infection, when she was in her prime. In fact, pulmonary infection is a common disease of older sheep. According to Professor Ian Wilmut, Dolly was also found to have arthritis, a common disease in older sheep. The terminal structures of chromosomes in eukaryotic cells are called telomeres, which determine the number of times the cell can divide. Each time the cell divides, the telomeres shorten; when the telomeres are depleted, the cell will lose its ability to divide. Scientists discovered in 1998 that the telomeres of cloned sheep Dolly were shorter than those of average sheep, that is to say, the cells were in a more aged state. This means that the cloned sheep Dolly may have had a shorter life span than average sheep and may have been more susceptible to disease or cancer than other sheep. Scientists at that time believed that this might have been due to the somatic cell used for cloning Dolly from adult sheep, which has the properties of adult cells, but this explanation was challenged by the results of the cloned cattle research.

Robert Lanza and others from Massachusetts, USA, used cattle aging cells to breed cloned cattle and obtained a total of six calves. When the calves were 5–10 months old, it was found that the telomeres of these calves were longer than those of ordinary calves of the same age, and some of them were even longer than those of normal newborn calves. This is quite different from the case of cloned sheep Dolly. At present, scientists cannot reasonably explain this phenomenon. However, the experiment suggests the possibility that in some cases, the cloning process may alter the molecular clock of mature cells and make them rejuvenate.

In other cases, however, the fate of the cloned cattle is regrettable. In Japan, some well-bred cloned cattle survived less than two months after birth. As of February 2000, a total of 121 somatic cell cloned cattle were born in Japan, while only 64 survived.

Scientific research shows that some of the cloned cattle have imperfect placental function; the oxygen content and growth factor concentration in blood are low; the thymus, spleen, and lymph glands of some cloned cattle have not developed normally; and the fetuses of cloned animals generally develop faster than normal animals. These are all possible causes of death.

Cloned cattle in France have similar problems. On May 14th, 2000, at the 20th Symposium on Biological Development, scientists from the French National Institute of Agricultural Research revealed that the cattle bred in a similar way as Dolly the sheep appeared as giants or had deformity syndrome during their growth and development.

In 1999, according to the famous medical journal *The Lancet*, the cloned cattle bred by the French National Academy of Agricultural Sciences died of severe anemia only 51 days after birth. After a year of research, scientists finally found that the success rate of somatic cell cloning is much lower than that of embryonic cell cloning. At that time, only 14 cloned cattle grew normally, and the other 150 miscarried during the embryonic stage or died soon after birth. The research results show that the main manifestation of the syndrome was fetal macrosomia. The average birth weight of normal cattle was 45 kilograms, while that of cloned cattle was more than 60 kilograms. Their abnormal development causes the pregnant cows to lose their appetite. Scientists had to take measures to abort or slaughter the cows. Many of the cloned calves that were lucky to be born soon died of cardiac abnormalities, uremia, or dyspnea accompanied by the inability to eat. This had not happened by accident, because the same is true of other similar experiments in the world. The mortality rate of cloned cattle has reached 70%.

The same problems exist in Chinese cloned goats. According to the Institute of Biotechnology of Northwest A&F University, the adult somatic cell cloned goat Yuanyuan, born on June 16, 2000, in the university's breeding farm, suffered from dyspnea and died of respiratory failure due to lung developmental defects. Professor Yong Zhang, who is responsible for the research on the cloned goat project at the Northwest A&F

University, recalled that shortly after the birth of Yuanyuan, the researchers conducted a detailed examination of its physical condition and found
that the breathing of Yuanyuan was a little difficult compared with an
ordinary lamb. At that time, they judged that this might be caused by
abnormal lung development, and Yuanyuan was observed in a close way.
But, one morning, Yuanyuan died of respiratory failure and had only lived
for 36 hours. Coincidentally, some of the transgenic goats cloned by
Yangzhou University have also died soon after birth.

This seems to indicate that the short lives of cloned animals are by no
means accidental. In addition to the need for improvement of cloning
techniques, there may also be some essential differences between
somatic cell cloning and embryonic development. Otherwise, in the
long-term evolution process, mammals will not choose the sexual
propagation way to produce offspring, and the asexual cloning method
has not become a way for higher animals to breed, which may have its
inherent defects.

After the cloned sheep Dolly came out, some people predicted that
cloning technology would have great commercial applications, such as
cloning the best racing horse, pet dog, and even man himself. However,
with the development of cloning technology, scientists increasingly doubt
that cloning has the same ability as a copying machine.

However, the fact is that four lambs have been cloned from the Roslin
Institute in Edinburgh, Britain, and they have shown obvious differences
in physical characteristics and behavior habits after they grew up. If the
cloning technology can achieve the theoretical effect, the four lambs
should look like that they came out of the same mold. "They do look similar, because the same breed of sheep will always be somewhat similar, but
you will never think that they are the cloned animals with identical genes."
Professor Campbell, who was involved in the reproduction of the four
lambs, said that as they grow older, they will become more and more different from each other.

Some scientists have analyzed that the reason for this difference may be
that the four cell nuclei used to clone the four lambs were placed into the
egg cells that were extracted from different ewes. Although the cell nucleus
is removed from each egg cell, the mitochondria in the cytoplasm still
contain a small amount of genetic material DNA, which is equivalent to

about 3% of the DNA in the cell nucleus. The mitochondrial DNA interacts with the cell nuclear DNA, affecting gene activity and embryonic development. Because of the different sources of egg cells, the way in which interaction occurs will be slightly different, resulting in cloned animals showing individual differences in their adulthood. This difference determines the viability and development ability of the embryo and may determine the physical characteristics of the male lambs when they come into adult.

Some scientists have also pointed out that gene mutation can also cause individual differences in cloned animals. Gene mutation means that the cell fails to replicate itself as it is when it divides and proliferates. If this genetic mutation occurs during embryonic development, it may affect all cells that have been divided after the mutation happened, and eventually the appearance of adult animals will change slightly.

The expression of genes in cells will also be affected by environmental factors. Studies have shown that the genes that control animal growth or behavioral habits are regulated by a number of switching systems that would turn on or off automatically at the appropriate time. In the case of humans, although these switches can control growth and development or the beginning of puberty, they themselves can also be affected by environmental factors, and the environmental conditions of cloned animals with the same genes may be different.

Scientists at the Roslin Institute believe that their research shows that, on the whole, it is impossible to revive a dead person or animal by using the technology that created Dolly the sheep. Professor Campbell said, "The only true clone is the identical twins, and anyone who is familiar with twins knows that, even twins have different characteristics and personalities."

People should have a correct understanding of cloning technology and abandon the false impression that science fiction has left on people that animal or human cloning is exact reproduction.

In addition to the inherent physiological or immunological defects of cloned animals, cloning techniques are immature, regardless of theory or method. In theory, there are still many questions that need to be clarified, such as the following: while all or most of the genes in the differentiated somatic cell nucleus have been switched off, how does the somatic cell restore its totipotency; whether the cloned animal will remember the age of the donor cell; whether the successive progeny of the cloned animal will accumulate the mutated gene; and what is the genetic role of cytoplasmic mitochondria in the process of cloning. In terms of methods, the success rate of cloning animals is still very low. In the experiment of cloning Dolly, the Wilmut research team fused a total of 277 nucleus-transferred egg cells, but only one living sheep Dolly was obtained. The success rate is only 0.36%. At the same time, the success rate of fetal fibroblasts and embryonic cells was only 1.7% and 1.1%, respectively. Even the Honolulu technology, with a less differentiated cumulus cell as the nucleus donor, only has a success rate of a few percentage points.

The inherent defects of cloning technology indicate that understanding of the laws of biological inheritance needs to be deepened. Biology is the most complex form of material existence in nature. Its wonder lies in its ability to inherit and mutate. The so-called heredity is that "Plant melons and get melons, sow beans and get beans." It is because of heredity that the species can continue to exist. But, organisms need to evolve and develop, which requires variation. The so-called variation is such that "A mother gives birth to nine children, and the children are different." Only in this way can it be ensured that the species will not decline and will not be eliminated by nature.

Even the cloned animals cannot change the natural laws of the heredity and variation of organisms; what we can do is to recognize and follow.

5. Why Human Cloning Should Be Prohibited

The cloning of human beings is prohibited

The invention of animal cloning technology has indeed brought practical benefits to human beings. It can save rare endangered animal species, and can also cultivate a large number of purely genetically identical purebred animals, so as to expand the fecundity of elite animal breeds. Scientists can transfer rare drug gene into the cloned animal, allowing the cloned animal to become a pharmaceutical factory to produce biological drugs in large quantities. The cloned transgenic pigs can also be used for organ transplantation to alleviate the current tight supply of transplanted organs.

The cloning technique became the Holy Flame of Prometheus, which let people see the beautiful prospects brought about by biological civilization. However, the cloning technique is still a "double-edged sword". The leaders of criminal syndicates, the bosses of drug cartels, and some dictators can also use cloning techniques to copy themselves. This is too terrible.

As a result, many people (including some scientists) regard cloning as a monster and Pandora's magic box. Once this box is opened, the disaster will be unprecedented.

The negative effects of human cloning involve ethics, morality, religion, law, and many aspects, which have attracted extensive attention of scientists, politicians, sociologists, anthropologists, jurists, policy analysts, ethicists, and relevant international organizations.

In terms of ethics, if a cloned person is bred from the somatic cell taken from an original person, is the cloned person the original person himself or the offspring of the original person? Of course, the purpose of cloning is nothing more than trying to copy another identical self, but it can't come true actually. The cloned person is actually much younger than the original person whose cell is used for cloning because the former develops late, and because the space, time, and environment of the cloned person and the original person who is cloned cannot be exactly the same; although they have the same genetic genes, there are still differences in personality, temperament, knowledge, thinking, and so on. Although the cloned person and the original person are similar in appearance, they are not the identical person. How does the cloned person rank in the family of the original person, how to continue the genealogy, how to get on the household registration? All of these will cause great ethical confusion, causing great trouble to society and daily life. If the cloned person gets married and has children, the hierarchy will be even more chaotic. The ethical concept that has been formed in the human heart since ancient times is bound to collapse.

In terms of morality, since the cloned person is a replica of the original person, they should also have the same rights and obligations. For example, the wife of the original person should also be the wife of the cloned person, and the child of the original person should also be the child of the cloned person. However, even if the original person is generous and agrees to his cloned person sharing his wife, how can the wife of the original person accept this inherited marriage without hard feelings? This is obviously contrary to the general social moral standards. In fact, this problem also exists between the children of the original person and the cloned person.

In terms of religion, the westerners are more concerned about it. For thousands of devout believers, religion undoubtedly has supreme power. They think that it is God who created human beings and all things, and you are undoubtedly trying to replace God by cloning human beings. This will offend the God, who will make disasters for human beings. Of course, religion itself sometimes goes against the modern sciences.

In terms of law, how do the original person and the cloned person inherit the heritage? If it is not done well, there will be trouble. If the original person commits a crime and lets the cloned person become a scapegoat, it is inevitable that the purpose of punishing the wicked and protecting the good people will not be obtained. If a criminal knows that he can regenerate through cloning after he has died, he will certainly take risks at all costs and the harm to society will be enormous.

Of course, there are other negative effects of human cloning. Human beings are the most advanced creatures in nature. It is obviously beneath human dignity for human beings to create clones that look like themselves but are listed outside human beings and can be enslaved at will. After long-term evolution, human beings have separated from the animal world and formed human society. Human genetic materials are the common wealth of all mankind. Human beings have the responsibility to protect their safety and not allow arbitrary modification of the genome, and human cloning is not conducive to the safety of human genetic materials. Humanity, personality, and family concept will be impacted, which is a serious challenge to people's personality, uncertainty, and interconnectedness. Human cloning can easily cause confusion in cultural concepts, customs, and habits. In all of these, at least from now on, the promotion of human cloning will cause great confusion in the family and society. And, if it is not done properly, it will breed crime and cause great retrogression in human civilization.

To this end, the emergence of human cloning has just begun; many international organizations and governments of some countries have enacted relevant laws, and have shown great determination in prohibiting the study of human cloning.

From May 8 to 10, 1997, the World Medical Association convened a meeting to call on all doctors and other researchers engaged in scientific research to spontaneously stay away from human cloning research.

On May 14, 1997, the resolution of the World Health Organization's 50th World Health Assembly on human cloning asserted that the use of cloning techniques to copy human individuals is ethically unacceptable and violates human dignity and morality.

In June 1997, the National Bioethics Advisory Commission advised the President to immediately request all companies, clinicians, researchers, and professional associations of private and non-federal funding departments to comply with the federal government's decision to suspend the study of human cloning. Professional and scientific associations should clarify their position that any attempt to create children with somatic cell nucleus transfer cloning is irresponsible, unethical, and against professional ethics.

From June 14 to 17, 1997, the Muslim Medical Organization recommended at a seminar that third parties, whether uterus, egg, sperm, or cloned cells, should not be allowed to be introduced into the relationship between husband and wife.

On June 22, 1997, at the summit meeting held in Denver, the United States, the Group of Seven Countries, and Russia issued a joint statement, "we agree that appropriate domestic measures and close international cooperation are needed to prohibit the use of somatic cell nucleus transfer to produce children."

In July 1997, the International Federation of Obstetricians and Gynecologists decided that the cloning of human individuals, whether by cell nucleus transfer or embryonic division, was unacceptable.

In March 1997, Minister Minzhang Chen of the China's Ministry of Health clearly stated his position on human cloning as "four no policies", that is, no approval, no support, no permission, and no acceptance.

On January 12, 1998, 19 European countries signed the "Agreement on the Prohibition of Cloning Humans" in Paris, France, prohibiting the use of any technology to create a cloned person with similar genes to any living or dead person. This is the first legal document in the world to ban human cloning.

At the end of 1998, two professors from the Affiliated Hospital of Kyung Hee University in South Korea successfully used a woman's egg cell to produce a germ that could give birth to a new life. But, because they could not pass the moral and ethical barriers, they had to stop the experiment.

On February 18, 2005, the 59th session of the UN General Assembly Law Committee passed a political declaration in the form of a resolution with the vote result of 71 votes in favor to 35 votes against, with 43 abstentions, requesting all countries to ban any form of human cloning that violates human dignity.

In fact, as early as 1992, when a scientific group of the World Health Organization reviewed medical assisted reproductive technology and related ethical issues, it emphasized that "extreme experimental forms, such as human cloning, interspecific insemination, making monsters and changing the genome of germ cells" must be prohibited.

On November 8, 1999, at the Tokyo Agricultural University, Japan, the human cell nucleus was transferred into the bovine unfertilized egg. Therefore, the cloning sub-committee of the Science and Technology Conference of the Japanese Prime Minister's Advisory Body compiled a final report and submitted it to the Bioethics Committee, advocating that human cloning should be prohibited by law. The main contents of the report include the following: it is forbidden to transfer human cloned embryos to human or animal mothers; it should also be forbidden to implant human cell nucleus into unfertilized egg of animals other than human beings so that they can develop into human cloned embryos in the mother's body, and carry on the research on the production of organs for transplantation with such human cloned embryos; the cell culture and tissue culture of non-human individuals can be unlimited; and the production of chimeric or mixed individual of human and animal has exceeded the cloned human individual, so it should be completely prohibited. After the implementation of the law, anyone who violates the law can be sentenced to imprisonment. Soon, the Japan Science and Technology Agency also submitted a bill banning human cloning to the Congress.

On September 29, 2000, the Japanese government announced again that it would impose a penalty of imprisonment of no more than 10 years and a fine of less than 10 million yen on the act of human cloning. At the same time, it abolished the bill on banning human cloning that had been passed by the parliament before, with imprisonment of less than 5 years and fine of less than 5 million yen. The Japanese government has been banning human cloning, as well as all the cloning of transplantation between the cells of humans and animals.

Human embryonic stem cell research has also been restricted in many countries. Germany implements the embryo protection law and strictly prohibits the study of human cloning and human embryos. The officials believe that the reasons for cloning human embryonic cells for medical purposes are not sufficient, and it is necessary to carefully weigh the advantages and disadvantages. Some people in Germany oppose the current embryo protection law and believe that it is necessary to modify it to meet the needs of modern medical development. They advocate the study of the cloning of human embryonic cells for medical purposes by a small number of research centers under strict supervision of the state. The United States prohibits the use of federal funds for such research, but there is no restriction on private funds. In November 1998, scientists from the University of Wisconsin and other institutions published a report in the American journal *Science* that they have successfully isolated and cultivated embryonic stem cells from human embryonic tissues. They can continue to grow and proliferate *in vitro* and have strong differentiation potential, making the United States take the lead in this field of research in the world. This breakthrough has attracted the attention of scientists from all over the world, and also triggered fierce disputes in ethics, morality, religion, law, and so on.

Despite all this, the emergence of cloned humans is still difficult to avoid. According to Professor Ian Wilmut, if a research team is ready to do this work and has obtained 1,000 human oocytes, it will be technically possible to create a cloned human in a year or two. According to a survey released by the Independent newspaper of Britain recently, despite the current general opposition to cloning research, many famous medical scientists still believe that the first cloned human baby will definitely appear in the future. Although most of the scientists interviewed said that they are not in favor of human cloning studies, they still believe that if technical and safety issues could be resolved, the reproductive cloning research for the purpose of human cloning might be carried out in the future.

Many scientists believe that although many countries in the world explicitly prohibit cloning research, there are always people in the world who are trying to do this research in some places. Due to the relatively simple and cheap instruments and equipment needed for human cloning research. It will happen regardless of whether the government approves it or not; it is difficult to stop.

However, if one day a cloned person really appears in the world, there is no need for people to panic. Even if they are cloned from the cells of criminals, they may not be villains when they grow up, because the character and quality of humans are not only determined by genetic factors. For example, the phenomenon of identical twins in humans is a good example. Twins from the same egg have identical genetic materials, but according to observation, their physiology, psychology, and personality are different. This is due to internal causes that are not exactly the same, as well as the influence of post-natal rearing conditions and social environment.

6. Therapeutic Cloning May Be Allowed

For a long time, the shortage of transplanted organs has been a major problem that has plagued the medical field. According to the laws of nature, most of the dead in the world are elderly people, cancer and heart disease patients, and their organs cannot be used for human transplantation due to defects. It is very difficult to find young and healthy organs, so the number of patients waiting for transplantation is rising. As early as 2000, there were hundreds of thousands of patients waiting for transplant organs worldwide, and only in the United States there were 62,000 people waiting for organs such as a new heart, lung, liver, and kidney. On August 22, 2015, China's first "Guide to Organ Donation in China" was released.

The chief editor of the guide, Jiefu Huang, head of the China National Organ Donation and Transplantation Committee, said, "At present, about 300,000 patients in China are waiting for organ transplants every year due to organ function failure, but there are only 10,000 organ transplants each year, and many patients are still waiting for transplants."

In this context, although human cell cloning for reproduction purposes has been widely condemned and banned by governments, human cell cloning for therapeutic purposes has gained certain understanding and support.

Lord Sainsbury, an official of the UK Department of Trade and Industry, has stated that he personally believes that human embryo cloning research should be allowed to develop human tissues for the treatment of diseases. Since American scholars have discovered stem cells that can control the development of human organ tissues, British scientists have suggested that human embryos should be cloned for extracting stem cells to regenerate human tissues. The British government ministers have not made a final decision on the matter because of fear of public opposition. At a meeting of the Bioindustry Association, Lord Sainsbury said that if the UK wanted to control the growing number of elderly people suffering from diseases such as Alzheimer's disease and Parkinson's disease, it was necessary to study human embryonic cloning. He believes that cloning therapy has brought hope to the treatment of diseases, and it is very likely to solve the problems of the human life quality decline caused by diseases.

On August 16, 2000, the British government finally announced that it would approve human embryo cloning experiments for therapeutic research purposes. As soon as the news came out, the international bio-medical community was immediately shocked. The British newspaper "The Independent" also published a photo of an embryo on the front page, saying that this six-day-old embryo heralds a bright future for life sciences in the 21st century. Some even optimistically predicted that human embryo cloning technology will lead to revolutionary changes in the treatment of some major human diseases. If any organ of the human body fails, it will be replaced by the cloned organ. Everything is as convenient as repairing a bicycle.

A report drafted by the British public health minister said that human embryo cloning experiments will open the way for finding new treatment options, with the aim of using young cells to grow a variety of human tissues to treat diseases that are impossible to be cured now. On August 23, 2000, then US President Bill Clinton also announced that he agreed to use federal funds to conduct research on cloning human embryos. He said that the US government made this decision after a careful review of the guidelines issued by the National Institutes of Health, because human embryo cloning research will bring "incredible potential benefits".

In this way, "therapeutic cloning" has gained some tolerance.

What is "therapeutic cloning"? Before the advent of somatic cell cloning technology, scientists could only obtain stem cells with strong dividing ability from the aborted, stillborn, or artificially inseminated human embryos for research. The advent of the cloned sheep "Dolly" means that human embryos can be cloned by using somatic cells, which will make the acquisition of stem cells easier. The doctor can remove some somatic cells from the patient for cloning, and let the formed blastocysts develop for 6–7 days, then extract the stem cells, and make them develop into cells, tissues, or organs, whose genetic characteristics are completely consistent with the patients. These tissues or organs are then transplanted to patients who provide the original cells, which is called "therapeutic cloning".

Clonal therapy is considered to have great potential for treating diseases of some human tissues, such as the brain, heart, and liver. It can be done in several steps. First, the nucleus is extracted from the patient's cells and combined with the empty human egg cell that the chromosomes have been removed from. After a few days, the synthesized cell develops into a cell ball with nearly 120 cells. In theory, if a cell ball is implanted into a human uterus, it can grow into a human clone of a patient, but this will be strictly prohibited in Britain. Therefore, according to the scientists' proposal, after obtaining the cell ball, stem cells should be extracted first, and then the cells of corresponding tissues, such as heart cells or brain neurons, should be cultured with stem cells as needed to replace the human tissues damaged by diseases.

So far, there have still been some problems with artificial organs; the organs that can be transplanted are extremely scarce and there will be rejection reactions. If the therapeutic cloning research is successful, patients will be able to easily obtain their own fully suitable transplant organs without any rejection reaction. At that time, blood cells, brain cells, bones, and internal organs will be replaced, which undoubtedly brings hope to patients with some diseases such as leukemia, Parkinson's disease, heart disease, and cancer.

PPL Biotech Co. in Britain attaches great importance to the research of "therapeutic cloning", on the one hand, because of the severe shortage of transplant organs in Britain, and, on the other hand, because organ transplantation can bring huge profits to the company.

Before launching human cell cloning for therapeutic purposes, PPL Biotech Co. first conducted a transitional pig cell cloning study. They believe that with the development of transgenic cloning technology, one day in the future, people may be able to use human cells to cultivate substitute transplant organs in the laboratory, but there is still a long way to go. In this situation, the only feasible way to solve the shortage of transplanted organs is to implement xenotransplantation, that is, the transplantation of cells, tissues, or organs of one animal into another animal.

In 2013, Matsunari reported a cloned pig with human pancreas which could be used for clinical transplantation.

The selection of pigs as breeding targets is mainly due to the fact that pig organs are physiologically closer to human organs, and pigs have the characteristics of strong fecundity, which provides the possibility of obtaining a large number of organs for transplantation. Scientists said that if a transplanted pig organ can continuously work for five years, it can basically solve the problem, because the organ can be cultivated in large quantities and can be surgically replaced when needed.

Some analysts believe that in the case of a global shortage of transplant organs, there will be a global market prospect of 6 billion US dollars of pig organs for human transplantation in the future. Of course, transplanting pig cells that can produce insulin also has the same market size. The huge market demand has injected a strong impetus into the research projects of PPL Biotech Co.

The development of cloned pigs with transgenic characteristics is an important step in the transplantation of xenogeneic organs, but far from achieving human transplantation of pig organs, many difficulties still have to be overcome. The first and most important one is the hyperacute rejection of the human body. Generally speaking, the rejection reaction of the human body is caused by leucocytes and antibodies attacking the foreign objects invading the body. However, xenotransplantation is different. When an organ has been implanted in the human body, before the leucocytes and antibodies act, it will be attacked by a complex of more than 20 enzymes in the human blood, which can coagulate the blood of the implanted organ within a few minutes, leading to hypoxia death.

For pigs, the main reason for this reaction is that pig vascular endothelial cells contain the saccharide molecules that humans do not have. When the pig organ is implanted into the human body, the human immune system will recognize these saccharide molecules as foreign objects and launch attacks, which can destroy the transplant organ within a few minutes. Therefore, to solve this problem, the strategy of PPL Biotech Co. is to modify the target gene responsible for the production of these saccharide molecules in pig cells to make them inactive, and then clone the modified cells. In this way, after the pig organs are transplanted into the human body, the hyperacute immune response of the human body can be avoided. At the same time, the company is also preparing to add a gene that produces natural protein into pigs to reduce the intensity of the immune response.

According to the reports of PPL Biotech Co., the saccharide molecules in pig organs are the main cause of immune rejection, but this is not the only reason. Therefore, the pig organs implanted in the human body may encounter other forms of rejection within 2–7 days after implantation. There are two main reasons for this rejection. One is caused by the loss of anticoagulants on the surface of human blood vessels. The anticoagulants have the effect of preventing blood clotting and blood vessel blocking, and these protective anticoagulants are lost when xenogeneic organs are transplanted. To this end, PPL Biotech Co. seeks to add a second gene to pigs so that when the organ is transplanted, the substitutes for anticoagulants can be produced when needed. The other reason is that the presence of excess vascular cell adhesion molecules (VCAM) on the surface of blood vessels also causes immune rejection. Usually, there are only a small number of such molecules in the human body, which act to induce leukocytes in the blood to penetrate into the site of infection and inflammation to resist the invasion of the bacteria. When a xenogeneic organ is transplanted, the VCAM molecules are overproduced, causing the transplanted organ to fail. To overcome this obstacle, PPL Biotech Co. plans to add a third gene to pigs. This gene allows a new protein to be produced inside the cells, capturing the VCAM molecules to protect the transplanted organs from being damaged. Scientists said that both of these products must be strictly controlled to ensure that they were produced in moderation only when they were needed, otherwise they would have catastrophic consequences.

The long-term immune rejection of the human body also needs to be overcome. In xenotransplantation, this immune rejection is mainly caused by the attack of T lymphocytes. There are many kinds of T lymphocytes, each of which recognizes a particular foreign invader and is part of the entire body's defense system. In order to prevent T lymphocytes from attacking the transplanted organs, PPL Biotech Co. intends to inject a small amount of modified pig cells into patients prior to organ transplantation. These cells can disable the recognition ability of T lymphocytes that are responsible for attacking the transplanted organs, while other T lymphocytes are not affected, which can still protect the human body from infection.

These are the main steps taken by PPL Biotech Co. to overcome the human immune response. If it goes well, it also needs to carry out primate animal experiments, safety studies to prevent swine virus infection, and human trials, and then achieve the purpose of clinical application.

However, the progress of the real studies on cultivation of the human cells for therapeutic cloning is relatively slow all over the world. Surprisingly, China has made substantial progress in the field of therapeutic cloning. Scientists have transferred the cell nucleus of patient's somatic cells into enucleated oocytes, and then developed them into blastocysts through a series of treatments, and achieved preliminary success in cloning. Therapeutic cloning has also been listed as a national key basic research project, which is divided into three parts: upper, middle, and lower reaches. Dr. Guoxiang Cheng from Shanghai Transgenic Research Center is responsible for the research of the upper reaches, Professor Huizhen Sheng from Shanghai Second Medical University and Professor Yilin Cao presided over the research of the middle and lower reaches, respectively. The overall goal is to transfer the cell nuclei of the patient's somatic cells into the enucleated oocytes, and then develop them into blastocysts after certain treatments, and then use the blastocysts to extract embryonic stem cells, and *in vitro* induce the differentiation of stem cells into specific tissues or organs, such as skin, cartilage, heart, liver, kidney, and bladder, and then transplant these tissues or organs to the patients. Using this method, the most difficult immune rejection during allogeneic organ transplantation will be fundamentally solved, and at the same time, a good and sufficient source of tissues or organs will be created.

Although therapeutic cloning has been welcomed by scientists, it has been opposed by some religious groups and individual governments.

On September 7, 2000, the European Parliament voted against the use of cloning technology for medical research with very close votes, declaring that medical cloning techniques, including the cloning of human embryos, would push medical research beyond the limits of responsibility. The European Parliament called on Britain to review its position on cloning technology and recommended that the United Nations should completely ban human cloning.

On August 11, 2004, Britain issued the world's first human embryo cloning license. The legal license is valid for one year and the embryo must be destroyed within 14 days. It is still illegal to breed cloned babies. Its main purpose is to increase human understanding of the development of its own embryos, increase human awareness of high-risk diseases, and promote human research on the treatment of high-risk diseases.

Why can an embryo within 14 days, that is, a pre-embryo, be used as research object? According to the results of a large number of embryological studies, 14 days is the final limit for the formation of twins, and before the formation of the external tissues of the embryo, namely, the ectoderm. What's more, the primitive streak has not yet appeared. Once the primitive streak appears, it means that embryonic cells have begun to develop and differentiate into various tissues and organs, showing their own characteristics. For example, it can develop into the spine, nervous system, etc. Therefore, the embryos before and after 14 days are essentially different. It is generally believed that the embryo at 14 days is still a cell mass without sensation and perception, which does not constitute the subject of morality, and the research on it does not violate human dignity.

James Watson, one of the discoverers of the DNA double helix, Nobel Prize winner, and famous molecular biologist, once said that it can be expected that many biologists, especially those engaged in asexual reproduction (cloning) research, will seriously consider its implications and launch scientific discussions to educate the people of the world.

The development of therapeutic cloning still has a long way to go. Scientists are continually exploring. It is expected that in the near future, cloned tissues or organs can be applied clinically, completely solving the problem of serious shortage of tissues and organs for transplantation.

7. Other Advances in Cloning Animals

The Indian bison is a kind of large wild animal inhabiting forests or bamboo forests in Southeast Asia and India. Today, there are only about 30,000 left in the world, which is already very rare and endangered.

The reason for its decreasing number is that in recent years, the wild habitat of the Indian bison has been shrinking, and poachers have been forced to take risks because of the high profits. Moreover, it is difficult for the Indian bison to successfully reproduce in zoos, and it is also the reason why this large wild animal family is not thriving.

For a long time, scientists hoped to breed this animal in large numbers to save this rare species, and the cloning technology really came in handy. On January 8, 2001, in Iowa, USA, scientists from a biotechnology company successfully removed the cell nuclei of cow egg cells and transferred the cell nuclei of Indian bison skin cells. Several months later, an Indian bison named Noah was born and weighed 80 pounds (about 36 kilograms). It was the world's first cloned bison, but it survived for only 1 day and died of dysentery.

Scientists said that after 12 hours of being born, the baby could walk without help and began to walk around instinctively. Noah's surrogate mother was a dairy cow. Although the baby was not of dairy cow family pedigree, nor was it a hybrid of a dairy cow and bison, but a 100% bison, she still showed selfless maternal love and loved little Noah. However, the surrogate mother was saddened that the child showed common symptoms of infection only one day after birth. Two days later, the cloned Indian bison Noah died of dysentery. Scientists tried their best to rescue it, but still failed to save the life of little Noah. The surrogate mother was safe and sound. Scientists believed that the cloned bison was infected with dysentery and it had nothing to do with the cloning technology itself.

Scientists are full of confidence in cloning technology, and hope it can save a lot of endangered wild animals in the world.

China also has plans to clone endangered wildlife. The white-flag dolphin is a large aquatic mammal living in the middle and lower reaches of the Yangtze River. It has been living on the earth for 20–30 million years. However, like giant pandas, the success rate of natural reproduction of the white-flag dolphin is extremely low. Coupled with factors such as the deterioration of the ecological environment of the Yangtze River, the number of white-flag dolphins living in the wild is continuously decreasing. There are less than one hundred in existence, which is only about one-tenth of the existing giant pandas. White-flag dolphin and giant panda are both known as "living fossils", belong to the 12 most endangered animals in the world, and are national first-class protected animals. In order to protect this rare and endangered species unique to China, scientists from the Wuhan Institute of Hydrobiology of the Chinese Academy of Sciences hope to clone it.

Xianfeng Zhang, a white-flag dolphin expert from the Wuhan Institute of Hydrobiology, said, "We have begun to extract somatic cells from the male white-flag dolphin named Qiqi that has been reared in captivity for 20 years, to obtain genetic information for cloning and other scientific experiments as soon as the necessary funds are available." The white-flag dolphin Qiqi raised by this institute is 2 meters long, weighs 125 kilograms, and is the only artificially reared white-flag dolphin.

The average life expectancy of the white-flag dolphin is about 30 years, and the 21-year-old Qiqi is about to enter its twilight years. Unfortunately, for various reasons, it has no offspring.

Professor Xianfeng Zhang said that in addition to the valuable genes extracted from Qiqi, they also obtained relevant information from the stranded, injured white-flag dolphins in the wild and established a somatic cell bank of white-flag dolphins to provide samples for white-flag dolphin cloning, reproductive physiology research, etc.

Animal protection experts believe that cloning is only an attempt to protect rare animals such as white-flag dolphins and giant pandas. It is urgent to improve their living environment in the wild and provide better breeding conditions.

According to an official from China's Ministry of Environmental Protection, as a major step in protecting the white-flag dolphin, China has invested more than 9.4 million yuan to build a white-flag dolphin conservation farm in Tongling, Anhui Province, which protects the white-flag dolphins, and the Yangtze river dolphins that are very similar in appearance and habits. Professor Xianfeng Zhang said that the white-flag dolphins in the Yangtze River are extremely rare. Their swimming speed is very fast. It is difficult for ordinary motor boats to catch up, and it is very difficult to change places to take care of them. However, the Wuhan Institute of Hydrobiology is seeking cooperation with other domestic institutions to raise about 1.5 million yuan to start research on white-flag dolphin cloning and other projects. On the contrary, scientists hope to use frozen cell technology to bring extinct wildlife back to life.

On September 1, 2015, Russian media reported that the curator of the Mammoth Museum in Russia said that the first laboratory for the cloning of extinct animals in Russia began to work in Yakutsk. The main task of the laboratory was to find living cells needed for subsequent cloning, so that mammoths could be regenerated. The report pointed out that in order to implement the project, the joint efforts of many scholars from different countries were gathered together. In order to get the cells, experts need not only to find well-preserved cells from animal remains in permafrost but also to find ways to thaw them normally. It was reported earlier that the long teeth of the "Red Mammoth" were excavated in the Nenets Autonomous Region.

Scientists at the Lucknow Institute of Paleobotany in India have discovered a fossil of male mosquito from 18 million years ago, and they hope to learn how mosquitoes and the fauna that now live in South Asia have evolved. It is reported that this fossil was found on a porcelain clay base in Kerala State, southwestern India. The mosquito was sealed in a small piece of resin and was well preserved. It is highly probable that the mosquito gene can be extracted from it. The scientists of the Lucknow Institute of Paleobotany, Arnold Plekas and Manoj Shukla, reported the findings to the Geological Society of India and requested the Hyderabad Research

Center for Cellular and Molecular Biology to extract and clone genes of the mosquito, and compare them with those of modern mosquitoes to understand the evolution of mosquitoes.

Previously, scientists at the institute also discovered a number of insect fossils of *Hemiptera*, *Hymenoptera*, and *Lepidoptera* wrapped in resin in a valley in Bihar state, northern India. These findings are extremely valuable for the research of gene evolution of insects. The discovery of these fossils is enough to show that these insects lived in the northern part of present-day India more than 10 million years ago, and most of the faunas living in the same area have evolved from then on. Therefore, the study of cloning of these extinct animals is of extraordinary significance.

In addition, the Spanish government has approved a plan to clone a newly extinct wild goat, and some scientists estimate that cloned dinosaurs will be born by 2030.

Transgenic animal cloning has attracted much attention. The success of somatic cell cloning has revolutionized the production of transgenic animals, and animal somatic cell cloning technology has provided technical possibilities for rapidly amplifying the germplasm creation effects of transgenic animals. The use of simple somatic cell transfection technology to transfer target genes can avoid the difficulty and inefficiency of livestock germ cell sources. At the same time, transgenic somatic cell lines can be used for the pre-checking of transgenic integration and the pre-selection of gender under laboratory conditions.

Before the cell nucleus transfer, the target fusion genes of foreign genes and marker genes (such as β -galactosidase gene *Lac Z* and neomycin resistance gene *Neo R*) are introduced into the cultured somatic cells, and the transgenic positive cells and their clones are screened by the expression of the marker genes. Then, the cell nucleus of positive cells is transferred into the enucleated oocytes, and the finally produced animals should theoretically be 100% positive transgenic animals. Using this method, scientists such as Schnieke have successfully obtained six transgenic sheep as early as 1997, of which three have human coagulation factor IX gene and marker gene, the neomycin resistance gene, three have marker genes only, and the integration rate of the foreign genes is over 50%. In the same year, Cibelli obtained three transgenic cattle using the cell nucleus transfer method, confirming the effectiveness of this method. It can be seen that one of the most important application directions of animal cloning technology today is the research and development of high value-added transgenic cloned animals.

On January 11, 2001, reports from the West Coast of the United States said that the first transgenic monkey bred by humans was born safely in the United States, which is the first transgenic primate in the world. Scientists reported in the *Science* journal published on January 12 of the same year that they added additional genes to the unfertilized oocytes of monkeys and successfully produced healthy and lively little monkey "Andi". Schatten, a scientist at the Oregon Regional Primate Research Center of Oregon Health Sciences University, was involved in this study, and said, "Andi was born on October 2, 2000, is very fit and is playing normally with his two companions."

According to the scientists, the gene added to Andi's body is only a simple marker gene, the purpose of which is to simply confirm its gene map, but the same transgenic method can make other experimental animals carry specific medical purpose genes.

Some people believe that this achievement may mean that the pace of human medical progress is accelerated, and the involved specific diseases may include diabetes, breast cancer, Parkinson's disease, and AIDS. For example, Schatten said, "We can simply introduce genes for Alzheimer's disease and speed up the development of vaccines against this disease." Because monkeys are more similar to humans than mice, we can get better answers to human diseases from a small number of animals, and accelerate the development of molecular medicine.

Schatten's research team reported the world's first cloned monkey from embryonic cells in 2000. Andi and the cloned monkey in early 2000 will quickly and safely help humans determine whether innovative therapies are safe and effective. Andi and his surrogate mother, as well as the female monkey "Tetra", cloned in 2000, are now healthy, Schatten said.

In the same period, great progress has been made in the cloning of transgenic animals in China. Three goats named "Lianlian", "Tiantian", and "Yunyun" appeared in the Shunyi Three-High-Tech Agricultural Experimental Demonstration Area, which are the first three transgenic goats with human α-antitrypsin gene in China. With the support of the Beijing Municipal Science and Technology Commission, China National Center for Biotechnology Development, and the Shunyi district government, these transgenic goats were co-bred by China Agricultural University and Beijing Xingluyuan Biotechnology Center. The modified human α-antitrypsin gene was introduced into a goat pronuclear stage embryo by microinjection. After more than two years of efforts, 1036 goat eggs were obtained and nearly 100 receptor ewes were transplanted. Two ewes and two rams were obtained. One ewe named Yingying died 3 days after birth.

According to the scientists, these three transgenic goats with human α-antitrypsin gene can produce more offspring through breeding, and one can extract specific drugs for chronic emphysema, congenital pulmonary fibrosis cyst, and other diseases from the milk of transgenic goats. In Britain, goat milk containing the drug costs 6000 US dollars a liter, and a ewe is like a natural pharmaceutical factory.

Researchers said that the pharmaceutical technology of transgenic animals has dozens of times the benefits of traditional animal cell culture technology, and a transgenic animal is a natural genetic medicine manufacturing factory. The success rate of the transgenic experiment reached 13.79%, which indicated the progress of transgenic technology in China. It explored another way for the production of biological drugs by using an animal mammary gland bioreactor, and filled the blank of the market of α-antitrypsin drugs in China.

In addition, the company PHP of the Netherlands has bred cattle that can secrete human lactoferrin, and the company LAS of Israel has bred sheep that can produce human serum albumin. The patent technology of Dolly the sheep bred by Roslin Institute in Edinburgh has also been transferred to the Therapeutic Branch of Australia Pacific Joint Stock Company in the form of license trade, which specializes in the production of milk therapeutic protein for transgenic livestock.

In recent years, following the transgenic animals, gene editing animals have appeared. Gene editing is to operate the target gene like text editing, but text editing is to add or delete words or punctuation, while gene editing is to knock out and add specific DNA fragments in the genome to purposely achieve the change of individual genes. Gene editing technology uses tools such as zinc finger nucleases (ZFN), transcription activator like effector nucleases (TALEN), and CRISPR/cas9. Among them, CRISPR/cas9 is a new generation of gene editor, which can be used to delete, add, activate, or inhibit other target genes, including the genes in the cells of a human, dog, zebrafish, bacteria, drosophila, mouse, yeast, nematode, crops, etc., which makes it easier to edit any gene, so it is a widely used biotechnology. The discoverers of CRISPR/cas9 are two great female scientists, Jennifer Doudna and Emmanuelle Charpentier, who won the 2015 Life Science Breakthrough Award.

In 2013, Professor Yong Zhang and others from the College of Veterinary Medicine, Northwest A & F University, developed an anti-mastitis dairy cow targeting lysostaphin gene by using the precise gene insertion technique mediated by zinc-finger nickase. In 2015, Professor Yong Zhang and others obtained 23 *Ipr1* gene targeting anti-tuberculosis cows by using Tale nickase-mediated gene precise editing technique. A related paper was published in *Proceedings of the National Academy of Sciences of the United States of America* (PNAS). The Journal of the American Academy of Sciences, founded in 1914, is one of the top comprehensive academic journals with the highest citation rate in the world.

Anti-mastitis dairy cows bred by Professor Yong Zhang and others by using the precise gene editing technique

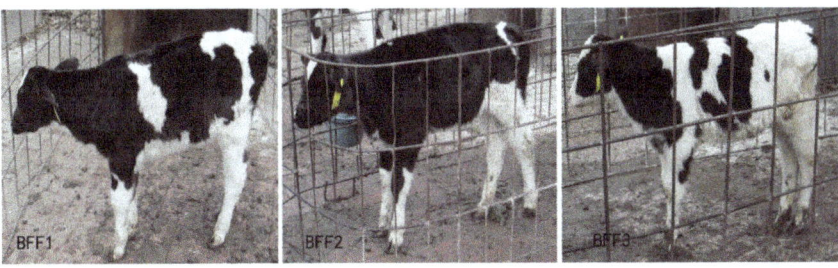

Anti-tuberculosis dairy cows bred by Professor Yong Zhang and others by using the precise gene editing technique

However, transgenic technology and cloning technology are still regarded as two important tools of life sciences. Based on the results of the deciphered genomes of many species, scientists have coordinated the use of transgenic and cloning technologies, and have been able to produce super organisms that human beings have never dared to imagine. This has brought a very bright dawn to many fields, such as medical treatment, traditional agriculture, and even industry.

Chapter 5

Cell Operations for Creating New Individuals

1. Breeding New Species by the Hybridizations of Cell Nuclei and Cytoplasms

The sheep Dolly is the world's first cloned animal using adult somatic cell, but it is not the first cloned animal. The earliest cloned animals were cloned by using embryonic cells. Because the potential of embryonic cells to develop into the whole organisms has long been a consensus in the biology community, the significance of this cloned animal cannot be compared with Dolly, but the cloning method provides a new idea for genetic breeding.

What was the earliest cloned animal? In 1952, American scientists Robert Briggs and Thomas King transferred the nucleus of early-stage tadpole embryonic cell into the enucleated frog egg, and the reconstituted cell successfully developed into a tadpole. The cloned tadpole is a replica of the original tadpole, triggering the first debate about cloning. The tadpole is the first cloned animal in the world, which rewrites the history of biotechnology.

In 1960 and 1962, scientists from the University of Oxford in Britain conducted cloning experiments on a kind of African toad with claws (*Xenopus laevis*). By artificially placing the nucleus of intestinal epithelial cells, hepatocytes, and kidney cells of the Xenopus tadpole into the egg

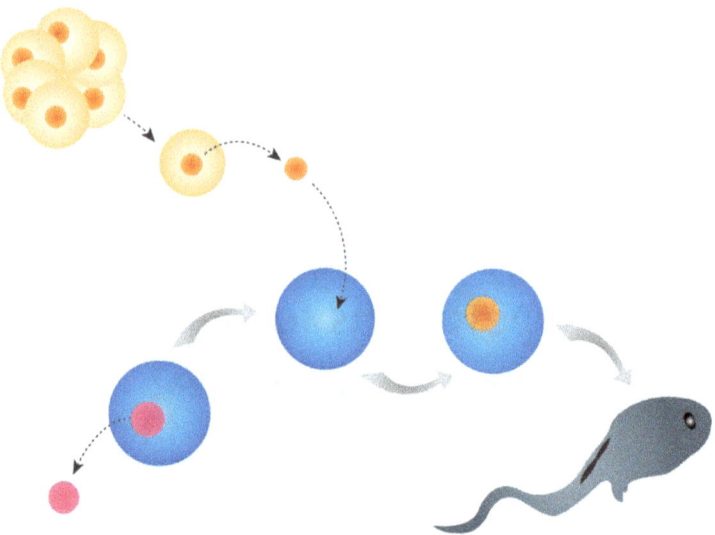

The cloned tadpole created by using cell nucleus transfer technique

cells whose nucleus was destroyed by the ultraviolet radiation, the Xenopus tadpoles grew up after great care.

In 1978, the famous Chinese biologist Dizhou Tong successfully carried out the cloning experiment of the black spotted frog (*Pelophylax nigromaculata*). He transferred the nuclei of red blood cells into the egg cells that had the nuclei previously removed. This kind of egg cells with changed nuclei finally grew into tadpoles that can swim freely in the water.

In the spring of 1979, scientists from the Institute of Hydrobiology of the Chinese Academy of Sciences used cells from the blastocyst stage of the crucian for artificial culture, and after 385 days and 59 generations of continuous subculture, the nuclei were extracted from the cultured cells under a microscope by using a glass tube with a diameter of about 10 microns, and then injected into the enucleated egg cells. Most of these nucleus-changed cells died in artificial culture. Only two of the nine hatched fries, and finally only one juvenile fish grew into an 8-centimeter-long crucian after more than 80 days of cultivation. These crucians do not undergo the combination of female and male cells, but only the egg cells change the nuclei of blastocyst cells. In fact, they are produced by the nucleus-changed eggs. They are cloned fishes.

The maturity of changing nucleus technology of fish and the success of changing nucleus of amphibians have excited a number of scientists engaged in the breeding of improved varieties. Since the nucleus of the crucian blastocyst cell can replace the nucleus of the crucian egg cell to obtain the cloned fish, can new hybrid fish be created by replacing the nucleus of different fish species?

Chinese scientists first raised this issue and solved it first. Researchers from the Institute of Hydrobiology of Chinese Academy of Sciences managed to replace the nuclei of crucian egg cells with the nuclei of carp embryo cells. After such recombination, the nuclei of carp embryo cells can actually be in peace with the cytoplasm of the crucian egg cells, and the processes of division and development are similar to that of the fertilized eggs. Finally, the bearded Carp-Crucians grow rapidly, which are completely like carps, but the numbers of lateral scales and vertebrae are the same as that of crucians, and the fish tastes as good as crucians.

In fact, as early as the 1960s, Dizhou Tong and his students began to study the nucleus transfer of fishes. They transferred the nuclei of blastocyst cells of one fish into the enucleated egg cells of another fish to study the development and traits of the nucleus-transferred fish.

These games of cell nucleus transport have been tried among animals of different species, genera, and subfamilies. The results showed that not only the nucleus-transferred egg cells could grow into small fish smoothly but also the several obtained hybrid fishes showed obvious heterosis. Since then, the nucleus transfer technology of fish cells has become an effective means of cultivating hybrid fish with excellent traits. At present, about the new economic fish species cultivated through nucleocytoplasmic hybridization technology, in addition to the aforementioned hybrid fish of carp cell nuclei and crucian cytoplasms, there are also hybrid fishes of grass carp cell nuclei and blunt-snout bream cytoplasms. They all show the characteristics of the two kinds of fishes and can be inherited to the next generation, which opens up a new way for the crossbreeding between distantly related animals.

But, it is also the use of embryonic cell nuclei for cloning. Why are cloned animals so different? Through careful analysis, it will be found that the cloned fishes or amphibians are changing the cell nuclei and cytoplasms of the same species, so the cloned animals and the animals that provide the nuclei are at least similar in appearance. Cloning using different cells of the same species can be used to propagate. However, the carp-crucian is different. It is the embryonic cell nucleus of the carp that develops in the egg cytoplasm of the crucian. Although the egg cells of the crucian have been enucleated, there is still genetic material DNA in the cytoplasmic mitochondria, and the genes contained in these DNAs are different from the genes of the carp. Therefore, crucian genes which are not found in carp cells are introduced into the recombined cells. On the other hand, during the development of fertilized eggs, the cell nucleus and cytoplasm interact, and the cytoplasm contains some substances that regulate the gene expression in the cell nucleus. Due to the different cytoplasmic compositions of carp and crucian, the regulation of genes in the cell nucleus must be different, resulting in different gene expressions. Thus, the carp-crucian has hybrid traits of two fishes.

In 2014, Yonghua Sun, director of the National Zebrafish Resource Center of the Chinese Academy of Sciences, reported the cross-species cloning of goldfish (*Carassius auratis*) and carp and the effects of cytoplasmic factors on the development of cloned fish. Although the surface characteristics such as long body, two pairs of whiskers, normal tail, and normal eyes of the cloned fish are similar to those of the common carp that provides cell nucleus, X-ray analysis shows that the number of vertebrae of cloned fish belongs to the range of goldfish (28–30 pieces), which is significantly different from the number of vertebrae of common carp (32–36 pieces). This indicates that cytoplasmic factors have an impact on the development of cloned fish, because there are also genetic materials (mitochondrial genes) in the cytoplasm.

Traditionally, cell nucleus transfer has been carried out by means of microscopic devices. In other words, the cells are sucked in by a special micropipette, and the cell membrane is broken by the pressure of the pipette wall, and then the cell nucleus is injected into the receptor cell together with a layer of cytoplasm wrapped around it. The receptor cell is mostly an activated animal egg cell, because the egg cell is large and easy to operate and the characteristics can be expressed through development. Scientists use this method to remove the cell nucleus of embryo obtained from the mating of grey mice, inject it into the fertilized egg cell of black mouse that is newly fertilized and enucleated, and transplant it into the uterus of a white mouse to continue to develop. Finally, a hybrid mouse with the cell nucleus of grey mouse and the cytoplasm of black mouse is born.

In 2004, Xinzhuang Wang and others reported the study of nucleocytoplasmic hybridization and early embryo development in mice and rabbits. The cells of the 2–8 cell stage mouse embryo were used as cell nuclei donors, and the rabbit oocytes were receptor cytoplasms, and electrofusion was used to form interspecific hybrid embryos.

In addition to cell nucleus transfer, can other organelles in the cell be transplanted? As early as the 1960s, someone transplanted the chloroplasts of spinach into the cytoplasms of cultured animal cells and obtained a kind of green hybrid cell of animal and plant. The cells are capable of normal division and the chloroplast structure remains intact.

We know that chloroplast has its own genetic system, and its DNA contains many genes. For example, chloroplast is responsible for photosynthesis of green plants. Carlson, an American scholar, was the first to engage in chloroplast transplantation. He cultured the protoplasts lacking chloroplast function together with the chloroplasts having normal function. As a result, complete green plants were regenerated from albino protoplasts containing alien chloroplasts. This is because the chloroplast enters the albino protoplast and functions normally, so green plants grow.

Transplanting chloroplasts from one plant to another can be used to improve plants with low photosynthetic efficiency. Scientists have found that some crops growing in tropical or subtropical regions, such as sugarcane and corn, have a more effective way to fix carbon dioxide, commonly known as the C4 pathway. Plants that carry out photosynthesis according to the C4 pathway are called C4 plants. Accordingly, the plants that carry out photosynthesis according to the C3 pathway are called C3 plants, such as rice, wheat, cotton, and soybean. It is found that four-carbon plants have higher photosynthetic efficiency than three-carbon plants, so the yield is higher.

Although green plants are the main fixers of solar energy conversion on the earth, the utilization rate of solar energy by crops is still very low, which is only 0.5–1.5% of the total solar energy put into the earth's surface by the sun, indicating that the potential of light energy utilization is huge. Scientists imagine that if the chloroplasts of plants with high photosynthetic efficiency are transferred to plants with low efficiency, it will have profound significance for food production all over the world.

In 2014, Zhenghan Qi and others reported that the big cyan rice, a new rice variety, was successfully bred by using nucleocytoplasmic hybridization technology. The cell nucleus of the big cyan rice comes from half of indica rice and japonica rice, and big cyan rice has the characteristics of low fertility or sterility, super high, delayed heading, etc.

2. Sphinx Chimeric Animal

The Sphinx in front of the Great Pyramid in Egypt is probably familiar to people. It is a huge stone statue composed of a human face and lion body. Why did the pharaoh of ancient Egypt build a monster in front of his mausoleum? It is probably a symbol of his power and majesty. Just think. The lion can be called the king of beasts, and man can still be superior to the lion. What a beautiful scene it is.

Sphinx (Egypt)

Actually, the Sphinx is not the exclusive right of the ancient Egyptians. There are also sphinxes in ancient Greek mythology, but the Sphinx of ancient Greece is winged and female, and the ancient Greeks invented a monster with a sheep head, lion body, and snake tail. In the myths of other countries and China, there are also monsters with a snake body and human face and mermaids with a fish body and human face. No matter what the original creators thought, in the eyes of the today's biologists, these monsters reflect the people's pursuit of creating life.

With the rapid development of life sciences, monsters such as the Sphinx are no longer absurd fantasies. Through the fusion of human cells and lion cells, it is theoretically possible to clone such monsters, but this is not allowed by law, ethics, morality, and religion, so it cannot be carried out. However, just as in plants, the fusion of potato cells and tomato cells can create two-story crops. The fusion between somatic cells of some economic animals may also create some new species of animals with better quality. Biologists refer to these animals with two or more species and their traits as chimeric animals.

How were the chimeric animals created? It is the aggregation of two or more early embryos or embryonic tissues that develops into animal individuals. If the embryonic cells of three different coat colors are gathered together, the chimeric animal will have three different colors of fur, and its body tissue is also composed of three kinds of embryonic components. These three kinds of embryonic components are coordinated in function, but each of them has its own characteristics. Each cell has its own genetic characteristics.

Taking mice as an example, there are two ways to breed chimeric animals: one is the aggregation method, which removes the zona pellucida from the embryos with protease or acid when the embryos have developed to an 8-cell stage. At this time, the adhesion of embryo cells increases. At 37°C, the two embryos are pressed gently by using the tweezers to make them stick together, or put the two embryos in a 4% agar well, so that the two embryos are in close contact. When cultured in a constant temperature incubator, the embryos can easily be wrapped together to form chimeric embryos. Next, because animals are heterotrophic (for example, they cannot produce organic nutrients by themselves, and they must ingest ready-made organic nutrients to maintain their survival), the embryos must be planted in the maternal uterus to obtain nutrition from the mother before they can grow and develop. Therefore, it is necessary to find a surrogate mother for the combined embryo and implant the embryo into its body so that it can grow and develop in the uterus. The other is the blastocyst injection method, which uses the mouse embryo of 4–5 days of pregnancy as the donor, takes out the inner cell mass, digests it into single cells with 0.25% trypsin, and then takes the receptor embryo of 3–5 days, puts it on the microoperation table, and injects the donor's inner cell mass into the blastocyst cavity of the receptor to make the cells close to the inner cell mass, and the cell surface of the inner cell mass is very sticky and easy to adhere. The chimeric embryos are then transferred to the uterus of the mouse as the surrogate mother.

Either of these two methods has advantages and disadvantages: the former is simple, but the chimeric effect is poor; the latter is complicated, but the chimeric effect is good. In addition, embryos used to prepare chimeric animals should have easily identifiable genetic markers, such as coat color, eye color, and ear shape.

In fact, the breeding of chimeric animals began in 1901, when Spemann cultivated frog chimera to study the development mechanisms of amphibian animals. In 1942, Nicholas and Hall tried to breed rat chimeras, but unfortunately, they did not succeed. By 1965, Mints was the first to successfully breed mouse chimeras. As a result, this technique attracted great attention from developmental biologists, and many scientists began to devote themselves to this research field. They fused the blastocyst cells of two different animals and obtained many chimeric animals. In these animals, both the somatic cells of each organ and germ cells contain two kinds of cells with different genetic characteristics, which can show the characteristics of the chimeric animals in the next generation and can also pass these characteristics to their offspring.

In 2014, Mikkers and others reported a method of breeding chimeric pigs with human pancreas. The cell nucleus of genetically modified pig somatic cell, which was different from germ cell, was first transplanted into the enucleated porcine oocyte and developed into a blastocyst *in vitro*. Human-induced pluripotent stem cells (iPS cells) were then transplanted into the blastocyst. The Human iPS cells are the stem cells that are genetically modified from human healthy somatic cells. Since the genetically modified pig somatic cells lack the ability to form specific organs (here, it is the pancreas), the development of specific organs can be accomplished by human iPS cells. Finally, the blastocyst was implanted into the surrogate sow. After the embryo matured, the born piglet was a chimeric pig with human organs [12].

China is at the advanced level in the world in the study of chimeric animals. In 1987, a research group led by researcher Deyu Lu from the Institute of Developmental Biology of the Chinese Academy of Sciences obtained three chimeric rabbits by the method of combining different kinds of rabbit embryos. In 1992, Beijing University successfully cultivated chimeric mice. In 1993, Northwest Agricultural University successfully bred chimeric goats. These have laid a good foundation for the future application of chimera technology in China for somatic cell breeding of pigs, cattle, and other livestock. In 1997, Yilin Cao, who was engaged in postdoctoral research at Harvard Medical School, successfully bred a mouse with human ear in Vacanti laboratory. It is a kind of auricle scaffold which is made of degradable material, pressed into ear shape, soaked in polylactic acid (PLA) solution to enhance its strength. Then, the cells were inoculated on the scaffold to allow them to grow. After 1–2 weeks of *in vitro* culture, a hole on the back of a nude mouse was cut and the "human ear" was implanted. Later, the scaffold will degrade and disappear, and the human ear will grow on the mouse's back. On December 26, 2011, Shanghai Science and Technology Museum officially opened, and the mouse with a human ear was listed in the exhibition area of Black Bamboo National High-Tech Industrial Development Zone.

On May 25, 2007, the world's first human–animal hybrid sheep was born from the work of a research team led by Professor Esmall Zanjani of the University of Nevada. This hybrid sheep containing 15% human cells cost the research team seven years. The purpose of this study was to solve the problem of transplanted organ shortage in the medical field by transferring human stem cells into animals and cultivating various organs suitable for transplantation.

In 2012, the *Grand Garden of Science* magazine reported on a magical double-sided cat called Venus, which not only had its own home page on Facebook but also has its own video clip on YouTube that has been accessed millions of times. At first glance of this 3-year-old tawny kitten, you will immediately understand why it is so popular. Half of its face is pure black and the eye is green, and the other half is a typical orange tabby and the eye is blue. The researchers took DNA samples of the different colors of skin on both sides of Venus, just as they did in a crime scene investigation. The results showed that the DNA extracted from one side of the skin was significantly different from that of the other side.

In 2014, Professor Nagashima of Japan successfully bred chimeric pigs. A white pig of No. 29 is covered with black pig hair, and more importantly has a black pig's pancreas inside. Originally, the research team injected the stem cells of a black pig into the embryo of a white pig, causing the genes carrying the instructions for developing the pancreas of animals in the white pig embryo to be turned off. The ultimate goal of this research is to grow human organs in pigs to meet the needs of those who need organ transplants. This could mean the end of waiting in the organ transplant list and eliminating the problem of organ rejection.

In 2016, Professor Qi Zhou of China and others reported that the embryonic stem cells of cynomolgus monkey were induced into pluripotent stem cells similar to the original state, and the chimeric blastocysts were formed after injecting into the host morula. The chimeric blastocysts were transferred into surrogate female monkeys and developed into chimeric fetuses. The results of analytical tests showed that the embryonic stem cells of cynomolgus monkey were involved in the differentiation and development of three germ layers and germ cells. This study provides a good primate model for researching the pluripotency of stem cells.

There are also chimeras of male and female, mainly existing in Insecta and Arachnoidea, such as chimeric crabs. The male blue crab has blue pincers and the female blue crab has red pincers. On May 21, 2005, a blue crab caught in the waters of Gwynn's Island, Virginia, USA, was noted as being unique, because it had a blue pincer and a red pincer. According to a crab expert at the Virginia Institute of Marine Sciences, this is a chimera of half male and half female. The last time people saw such a crab was in the waters of Smith Island in 1979. Other chimeric animals include chimeric lobsters, chimeric butterflies, chimeric spiders, chimeric stick insects, chimeric chickens, chimeric moths, and chimeric North American rosefinches, which are very few in number. The formation of these chimeric animals is mainly caused by external interferences or disorders of their own gene regulation during their development.

Chimeric animals, on the one hand, provide an ideal model for biologists to study embryonic development and the interaction between cells in the process of embryonic development and, on the other hand, also provide a new way for breeding excellent economic animal varieties or serving clinical practice.

3. Embryo Transfer, Breeding Elite Animals

A cow can only give birth to one calf a year, and can at most have about 10 calves in her lifetime. However, its ovaries contain as many as 75,000 eggs, so a large number of eggs are wasted. For a high-yield cow, if it is allowed to breed naturally, it is a pity that the breeding efficiency is so low. Is there any way to improve the reproductive rate of elite animals?

According to common sense, it is impossible for us to shorten the pregnancy time of domestic animals, and it is very difficult to achieve multiple births. Is there no other way to go? Of course, there is. The development of cell engineering provides us with new ideas.

It is well known that in the breeding process of elite animals, sperms must be provided by elite male animals, the elite female animals provide eggs, sperms and eggs are fertilized in the reproductive tract of the female animals, and then the fertilized eggs develop in the womb. In fact, all the genetic information that determines the development of an animal is contained in the fertilized eggs, and whether they are developed in the uterus of elite female animals, the offspring of the elite animals will still be elite animals. It is like a baby with black hair, black eyes, and yellow skin. No matter whether he was born in the United States or China, he can't change his blood line of black hair, black eyes, and yellow skin when he is an adult. In the same way, for the fertilized eggs or embryos of elite animals, even if they are developed in the wombs of mediocre female animals, the offspring that are produced are still elite animals.

If mediocre female animals are used as surrogate mothers to undertake the long embryo development process, while the elite animals only provide the fertile eggs or early embryos, the fecundity of the elite animals will be greatly improved.

This kind of elite female domestic animal breed reproduction strategy is called embryo transfer [13], also known as "breeding babies in borrowed wombs" and "fertilized egg transfer", which can be carried out across species, such as transferring bovine embryos into sheep. Using hormone treatment, the donor female domestic animals can discharge several or even a dozen eggs at a time, and the discharged eggs are fertilized to form embryos. After 6 to 7 days, the embryos are flushed out from the uteruses of the donor female domestic animals and transferred to the specific parts of genital tracts of the receptor female domestic animals with estrus synchronization, allowing them to continue to grow and develop until they grow into litters. Of course, the eggs can also be taken out from the cow body for *in vitro* fertilization. After being cultivated into embryos, they are sent to the cows to give birth to calves. Such calves, also known as test tube calves, are an important part of embryo transfer technology. In addition, embryo transfer is also the basis of embryo bioengineering, such as somatic cell cloning technology. After the cloned embryos are obtained in the laboratory, they must be transferred into the receptor animals using embryo transfer technology, and finally cloned animals are obtained.

The first scientist to carry out embryo transfer experiments was Walter Heape from the University of Cambridge. In 1890, he transplanted a 4-cell Angora rabbit embryo into a mated Belgian hare which successfully gave birth to two Angora rabbits. It was priceless to be able to invent such high technology at that time. In the 1930s and 1940s, scientists began to experiment on large domestic animals such as cows, sheep, pigs, and horses. In 1951, a calf was finally born in the United States through embryo transfer. Since then, embryo transfer technology has come out of the laboratory and created considerable economic benefits.

So far, bovine embryo transfer technology has played an important role in animal husbandry. In countries such as the United States, the increase in milk demand has stimulated the rapid growth of high-yielding dairy cows. Embryo transfer technology has become a widely used technology, bringing vitality to animal husbandry production in these countries. In the United States, 60% to 70% of cows are obtained through embryo transfer technology. Due to the widespread use of high-quality dairy cows, the total number of dairy cows in the United States has been reduced by more than half. At the same time, embryo transfer companies have sprung up all over the world like bamboo shoots after a spring rain. According to the statistics of the International Embryo Transfer Society (IETS) on global embryo transfer data in 2010, a total of 104,651 cows were subjected to flush embryos, 732,227 embryos were available, and 590,561 embryos were transferred, including 263,036 fresh embryos and 327,525 frozen embryos, which increased by 1%, 4%, 8%, 13%, and 11%, respectively, compared with 2009. Since the statistics of IETS have not yet covered all regions of the world, these data may be lower than the actual situation. It is concluded that embryo transfer technology has become one of the most dynamic and practical breeding techniques in cattle industry.

In addition to cattle, the embryo transfer technology in other animals has also been developed. In 1983, a graduate student studying at the University of Cambridge in the United Kingdom successfully transplanted four *in vitro* fertilized pig embryos and successfully obtained 4 piglets. This is the first time in the world to complete a very difficult pig embryo transfer. In 2013, Wei Liu and others reported the use of embryo transfer technology to breed SPF mice. This kind of mouse has become an internationally recognized standard laboratory animal, because it does not carry major potential infections and conditional pathogens and pathogens that interfere with scientific experiments. In 2014, Xiufang Jiang reported that the embryo transfer of goats was carried out among local farmers in Jiangsu Province, and the receptor sheep were all local leading breed Xuhuai white goats. A total of 56 goats were transferred and a 43% pregnancy rate was obtained. There are also scientists working on using embryo transfer technology to save rare animals that are on the verge of extinction. In 1985, a zebra was born using embryo transfer technology in the London Zoo for the first time. It was born from the belly of a common horse. In 1987, the National Zoo of the United States announced that using domestic cats as surrogate mothers, the fertilized eggs of a kind of rare endangered cat were transplanted, and three litters of precious kittens were successfully obtained. On June 5, 2014, the Xinhua News Agency reported that a purebred "Ferghana horse" was born in the Yili grassland of Tekes County, Xinjiang, China, but the mother of the newborn foal was just a common Ili horse, and has no blood relationship with it. Jun Sun, the secretary of the Party Committee and Deputy Director of the Tekes County Animal Husbandry and Veterinary Bureau, stated that 11 months ago, experts used embryo transfer technology to take out the embryos from the "Ferghana horse" and transplant them into the uteruses of six Ili horses, and cause them to be successfully pregnant.

China is close to the international advanced level in the embryo transfer of dairy cows. In 1988, the average number of available embryos obtained from each donor cow was 4, the average conception rate of fresh embryo transfer receptor cows was about 35%, and the conception rate of frozen embryo transfer cows was about 20%. Later, the embryo transfer technology of the Boer goat which is suitable for China to promote vigorously and has the reputation of being the world's best goat varieties, also achieved a breakthrough. Scientists form the Northwest A&F University used superovulation technology to flush out 46 available embryos from the uterus of a donor Boer goat and then successfully transplanted them into the uteruses of 23 receptor goats. According to Professor Zhongying Dou, the chief scientist of this project, this is the latest breakthrough in similar research and practices all over the world. If hormones are used for superovulation, 10 eggs can be discharged at a time, and 6 to 7 of them can become normal embryos. After being transplanted into the uterus of the surrogate mother, they can grow into the average 3 to 4 embryos. One ewe can super-ovulate five times a year, so that the number of litters born in one year can be increased by more than 15 times than in the natural case. In October 2000, Aershan Agriculture and Animal Husbandry Technology Co., Ltd., located in the Wulagai Development Zone of Xilin Gol League, Inner Mongolia Autonomous Region, China, implemented the world's first "10,000 high-quality breeding sheep embryo transfer project". In order to carry out this extra large-scale embryo transfer project, the two foreign-funded enterprises of Yichang Technology Group and CCI Group of the United States invested more than 100 million yuan in the first phase. A 50,000-square-meter embryo transfer center and breeding farm was established on the Inner Mongolia natural grassland of more than 100 square kilometers. On April 8, 2003, the *Guangming Daily* reported that China's first embryo breeding project base for elite sheep was established in Linxi County, Hebei Province, making Boer goat breeding a pillar industry in the country. In 2015, Xiangdong Kan reported the success of using the embryo transfer technology to breed the Suffolk mutton breeding sheep in Tibet, promoting the development of the local mutton sheep industry. In 2016, Zhoushan Cai and others reported the promotion of embryo transfer technology in 27 towns in Liangzhou District. With the simmental crossbreed cows as receptors, a total of 1796 frozen embryos of high-yield dairy cows were transplanted. 971 cows were pregnant, with an average pregnancy rate of 54.06%, and 952 calves were alive.

The Institute of Genetics of the Chinese Academy of Sciences cooperated with the Three-North Breeding Sheep Farm in Inner Mongolia to use the super-ovulation technique to make a 7-and-a-half-year-old black karakul sheep lay 18 eggs at a time. After fertilization, these eggs were transplanted into white Mongolian sheep and resulted in 11 karakul lambskin sheep. In recent years, the level of embryo transfer has been further improved. The average number of available embryos per donor cow is 5 to 9, and the average conception rate of fresh embryo transfer is about 50%. Not only that, China has also made some achievements in the transplantation of segmented embryos. By dividing the embryo into several pieces and then transplanting them separately, the fecundity of the elite livestock can be further expanded. The Northwest Agricultural University divided a bovine embryo into two parts, which had developed into two calves. The Institute of Genetics of the Chinese Academy of Sciences divided the bovine embryo into four pieces and developed them into four calves.

A working group of 15 bioengineering doctors, masters, and researchers led by Australian biology experts and Professor Zhongcheng Zhang from China Agricultural University, used the technology of "breeding babies in borrowed wombs" and advanced bioengineering methods such as embryo thawing. All 10,000 embryos were transplanted into the uteruses of different local sheep after 15 days and nights of hard work by Chinese and foreign experts. The whole project was successfully completed. The transplanted embryos were produced in the Australian laboratory by Chinese and Australian experts using Chinese test tube embryo technology. The transplanted embryo varieties include sheep breeds Suffolk, Poll Dorset, Texel, and Boer goats. They have the characteristics of large size, fast growth, high meat production rate, delicious meat, etc. The embryos of these high-quality sheep and goats will be bred in the bellies of local ewes and then born, thus quickly establishing an excellent population adapted to the local natural environment.

Embryo transfer technology not only can be applied to the rapid propagation of elite livestock but also has a place in animal breeding. For example, it can be used for the breeding of super mice and super pigs. The eggs or early-stage embryos of transgenic animals that have been genetically modified *in vitro* must be returned to the uterus to grow normally, so animal embryo transfer is also the indispensable part of animal genetic engineering breeding.

Through the embryo transfer technology, the infertile female livestock with genetic values can give birth to offspring, and the embryo gender can also be selected during the transfer. In addition, the embryo transfer of dairy cows and dairy sheep may create more economic value.

4. Test Tube Babies, Overcome Infertility

Since ancient times, China has had the saying of "the happiness of a family union". It is true that the family is happy when they get together. But, for families that don't have children, they don't enjoy this happiness.

The "Tell the Truth" program of China Central Television has reported that Xiurong Zhang and her husband Fujin Liu, who lived in Panjin, Liaoning province, China, originally had a happy family, but there was an unpredictable car accident, their only 24-year-old son was killed and it was a disaster for them. They want have another child, but Xiurong Zhang, the mother, had done bilateral tubal ligation and was already middle-aged, meaning that the couple could not rely on the method of natural conception to achieve the desire of having a child.

In December 1997, the couple came to a hospital in Shenyang City with a glimmer of hope. The doctor received the couple and expressed deep sympathy after listening to their experiences. After careful examination of them, the doctor found that both parties had good fertility conditions and advised them to accept scientific assisted reproductive technology to have a baby. The couple immediately agreed. On January 17, 1998, the doctor performed the surgery on Xiurong Zhang. Then, she was successfully pregnant and everything was normal during the pregnancy. On October 1 of the same year, Xiurong Zhang went into labor.

At 8 o'clock on the same day, Xiurong Zhang was sent to the delivery room for a cesarean section. It took 30 minutes from opening the abdomen to closing the abdomen. At 8:50, a chubby, very lovely healthy baby girl was born, whose hair was dark and thick, and eyes opened in a few minutes. She kept crying, and crying loudly. She didn't cry until the medical staff wrapped her up and put her next to her mother's face. At that time, 46-year-old Xiurong Zhang had been sterilized for 15 years. After losing her 24-year-old son, she could not help being a mother again.

The newly born baby girl weighed 3,900 grams and was 51 centimeters in length. She was not an ordinary baby born through natural pregnancy, but a test tube baby, coming to this world by *in vitro* fertilization and embryo transfer technology. One needs to give thanks to science that Xiurong Zhang became a mother again. Fujin Liu, the father, at nearly half a hundred years old, could hardly conceal his excitement. What made him happy was that his daughter's and his birthdays are on the same day. He happily named his daughter "Tongqing", which means celebrate together.

In April 1999, Xiurong Zhang and Fujin Liu, with their lively and lovely daughter "Tongqing", were invited to the "Tell the Truth" program of the China Central Television, which made hundreds of millions of viewers across the country see a real test tube baby. At that time, "Tongqing" was only six months old. Host Yongyuan Cui wrote some encouraging words for her: "The Whole World Is Peaceful" and "The Whole World Is Celebrating".

This is the story of Tongqing, the test tube baby reported by the media. In fact, many couples in the world have the similar experiences as the parents of Tongqing. In 2003, Ms. Li used test tube baby technology for the treatment of fallopian tube obstruction and polycystic ovary syndrome at the Reproductive Medicine Center of Tangdu Hospital in China. In August of that year, 12 eggs were successfully taken, 12 embryos formed, 2 transplanted, and 7 frozen. She got pregnant in that month successfully, and a healthy baby boy weighing 2900 grams was born the next year. The remaining 7 embryos were still kept in Tangdu Hospital. In 2015, Ms. Li wanted to have a second child through frozen-thawed embryo transfer technology. After a series of examinations, her physical condition was found suitable for receiving test tube baby technology again. After the embryo thawing and resuscitation with newly prepared medium, 3 of the 7 embryos survived. The doctors chose two of them for transplantation and she was successfully pregnant again. February 24, 2016, 12 years later, Ms. Li, who is forty years old, once again gave birth to a healthy baby boy weighing 3440 grams by caesarean section.

According to statistics, there are currently 10% to 15% of couples of childbearing age suffering from infertility, which makes family relationships very tense. Of course, some infertility can be cured by drugs or surgery, but for some complicated infertility patients, drugs or surgery will not help. In addition, some couples who originally had fertility did not want children when they were young because they were busy with their careers or other reasons, but when they want to have baby, they found it was too late due to the lost fertility.

Modern test tube baby technology has developed rapidly and has reached the third generation. The generation of test tube baby technology is ranked according to the difficulty of the technology and the level of operation.

The first generation of test tube baby technology, also known as *in vitro* fertilization and embryo transfer technique, mainly solves female infertility, such as tubal ligation or partially obstructed. The father's sperms and mother's eggs are extracted by artificial methods, then fertilized *in vitro* to form an embryo, and transplanted into the mother's uterus for development. At the end of the 1980s, the first test tube baby in China was born in the Third Affiliated Hospital of Beijing Medical University.

The second generation of test tube baby technology, also known as intracytoplasmic sperm injection technology, mainly solves male infertility, for example, male azoospermia, severe oligospermia, asthenospermia, sperm deformity, and obstructive azoospermia. The process of sperms entering into eggs is done manually, which is much more difficult than the first generation. The technical difficulty lies mainly in the search for sperm. Some male patients have very few sperms. One usually needs to puncture the epididymis and testicles to find a few numbers of sperms that are hidden in red blood cells and nerve tissues, and then inject the sperms into the eggs by means of micromanipulation. The fertilized eggs are cultured for 6 days under suitable *in vitro* conditions and then transplanted into the uterine cavity of the mother. So far, seven couples in Jiangsu Province of China have obtained the second generation of test tube babies through this difficult technique.

The third generation of test tube baby technology, also known as pre-implantation genetic diagnosis, or embryo screening, mainly solves eugenics, and its technical difficulty is also the highest. According to statistics from scientists, there are more than 8,000 kinds of genetic diseases and chromosomal diseases that can be inherited from parents to their offspring, such as hemophilia and sickle cell anemia. In the past, these diseases could only be found through the puncture examinations of amniotic fluid and chorionic villus when women had been pregnant for more than 4 months. Once the diagnosis was confirmed, induced labor would cause great damage to women's physiology and psychology. By using the third generation of test tube baby technology, the eggs and sperms are taken out to be fertilized *in vitro* to form multiple embryos. When each embryo grows to more than 8 cells, 1–2 cells are taken from each embryo for chromosome or gene defect detection. The embryo without defect is implanted into the uterine cavity for further pregnancy. The test tube baby born in this way must be a healthy baby.

Since July 25, 1978, when the world's first test tube baby Louise Joy Brown was born in Britain, the inventor of the technology, Robert Geoffrey Edwards, known as the "father of test tube baby", won the 2010 Nobel Prize in physiology or medicine. After that, the cultivation of test tube babies sprang up like mushrooms in various countries. On March 10, 1988, Mengzhu Zheng, the first test tube baby in China, was born at the Third Clinical Medical College of Beijing Medical University (now the Third Affiliated Hospital of Beijing University). The test tube baby technology has been accepted by more and more people. It is reported that there are 150 hospitals that can provide technical services for test tube babies in Japan, 250 hospitals in the United States, and one in 10 children in Finland is a test tube baby. Test tube babies have been accepted by more and more people, which is nothing new now. The Tianjin Daily reported that more than 300,000 test tube babies have been born in the world. Long-term follow-up studies of these test tube babies showed that they were no differences from those born naturally.

Although test tube babies have become more and more popular, many people still have a shadow that cannot be easily erased. They think that the test tube baby is not their own child, so even if they have infertility, they are not willing to let others know that they want to have a test tube baby.

In fact, the eggs and sperms used to grow test tube babies come from the couple themselves. From a genetic point of view, test tube babies are absolutely their own, just like babies born in natural pregnancies.

The latest research of American scientists shows that test tube babies are likely to have congenital defects. It is a very important process to cultivate test tube babies, that is, a small hole is drilled on the early embryonic membrane by using laser or microneedle, so that the early embryo cultured in the test tube can be smoothly implanted into the uterus of woman, and then pregnancy can be caused. However, this method is very risky, because the congenital defect probability of twins born with this method is 1.7 times higher than that of normal twins.

The Lund Hospital in Sweden also found through tests that the IQ of test tube children is slightly lower than that of ordinary children. The hospital's pediatrician conducted an extensive IQ test on 72 test tube children born between 1986 and 1992 in the Skane region of southern Sweden. The results showed that the average IQ of these test tube children was 3.3% lower than that of normal children. Experts believe that one of the main reasons why test tube children's IQ is slightly lower than normal children is that they have a higher proportion of premature births than normal children. Among the 72 test tube children tested, the proportion of premature births was up to 30%. However, this test shows that test tube children also have many advantages, such as strong self-confidence and being able to please their parents. Therefore, experts believe that test tube children are completely the same as normal children.

The phenomenon of multiply pregnancy in test tube babies is very prominent. The incidence rate of twin and multiple births in natural insemination is much lower than that of single birth, and it is quite different compared with the incidence rate of single birth. The incidence rate of twin births and triplets is about 1:66 and 1:8000, and the incidence rate of more than triplets is even lower. But, in test tube babies, the incidence rate of twin births, triplets, and more than triplets is about 20.7%, 4%, and 0.4% respectively, which is caused by the use of follicle stimulating hormone (FSH), a kind of ovulation promoting drug. The more follicle stimulating hormone is used, the more mature follicles will be. In this way, multiple eggs can be taken for fertilization and multiple embryos can be transferred at one time. Although a large number of embryos are transferred each time, only one embryo survives in most cases due to the different viability of the embryo and the different environment of the uterus. When the several transplanted embryos have strong vitality and the endometrial development is well, multiple births will be formed. Experts believe that if the multiple birth phenomenon of test tube babies is serious, it will not only be harmful to control the quantity of the population and improve the quality of the population, but also bring heavy financial burden to the family.

There are more males and fewer females in test tube babies, so there is a gender imbalance. According to the *Guangming Daily* reported on March 22, 2016, the research team of Professor Jianhui Tian of China Agricultural University found that the *in vitro* fertilization embryos of mice had insufficient X chromosome inactivation, and it was concluded that this may be the main cause of gender imbalance in test tube babies.

As the proportion of male infertility and female infertility in China is increasing year by year, more and more couples will choose test tube baby technology to get their own baby [14]. However, we must avoid multiple births and gender imbalance as far as possible, so as to facilitate eugenics and improve the population quality of the whole society.

5. Artificial Seeds, Better Breeding

The rapid asexual reproduction technology using plant tissues such as leaf tip and stem tip in test tubes has the advantages of fast propagation speed, large production capacity avoidance of plant virus infection, no restrictions on seasonal and environmental conditions, and suitability for industrial production. Therefore, it has been widely used in production and has achieved considerable economic benefits. However, the production of such test tube plantlets also has some technical defects; it needs a series of complicated processes, such as induction of rooting, transplanting, packaging, storage, and transportation, in order to play a role in production. In contrast, artificial seeds have many advantages in this regard.

What are artificial seeds? Artificial seeds are also known as "synthetic seeds" and "somatic cell seeds". The technology of artificial seeds is relatively young and practical in cell engineering. It was first proposed by the

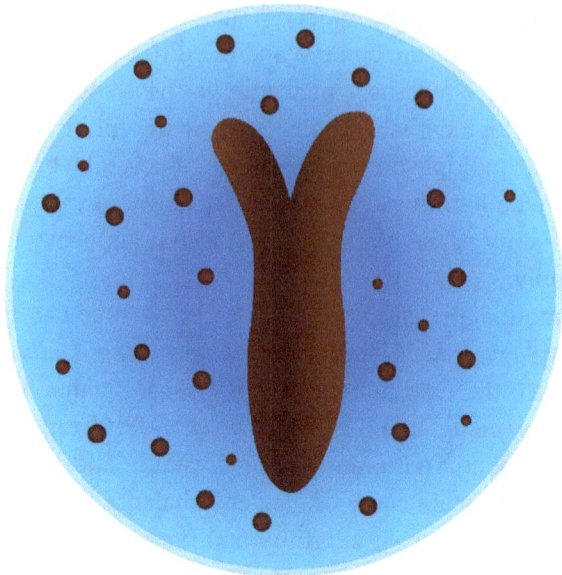

Artificial seed

British scientist Murashige at the 4th International Conference on Plant Tissue and Cell Culture technologies in 1978, and attracted widespread attention in the academic community. He believed that by utilizing the characteristics of somatic embryogenesis, somatic embryos can be embedded in capsules to form artificial seeds with properties of seeds and can be sowed directly in the field.

This idea aroused great interest. In 1985, the Japanese scholar Kamada first extended the concept of artificial seeds and believed that the meristems such as buds, callus, embryoids, and growth points that are obtained from tissue culture and have the abilities of developing into complete plants can be embedded with appropriate methods to form the artificial seed granules that have the abilities of replacing natural seeds for sowing. In 1986, Redenbaugh and others successfully used sodium alginate to embed single somatic embryos to produce artificial seeds. In 1998, Chinese scientist Zhenghua Chen and others further extended the concept of artificial seeds to the following: any kind of propagule, whether embedded in film capsule, exposed, or dried, can be called artificial seeds as long as it can develop into a complete plant. After that, artificial seeds gradually developed in China. In 2003, Lianzhen Yang reported on the preparation process of artificial banana seeds. In 2010, Aizhen Li and others reported the screening of artificial seed coating materials for carrot. In 2016, Baozhen Zhou reported on the artificial seed production process of *Dendrobium candidum*.

As we know, natural seeds are composed of seed coats, embryos, endosperms, and so on. Among them, embryos are the key components of seeds, which can germinate and produce the next generation. Endosperms provide nutrition only during embryo germination. In plant tissue culture, callus can be formed by a small piece of leaf tip, shoot tip, protoplast, etc., and can induce the formation of embryoids. The embryoids have polarity, that is to say, they have developmental potentials of buds and roots. Although the origin of embryoids is different from that of natural seed embryos, they have the same function. After being planted in the soil, the embryoids can germinate, take root, and grow into complete plants like natural seeds. Different from the germ cells in the seeds, the embryoids are developed from somatic cells, so they keep the excellent characters of the original varieties.

Since the embryoids that have the characteristics of somatic cells have the potential to regenerate complete plants, why not use them directly to simplify the process of tissue culture? So, scientists came up with a high-tech idea of using the embryoids to prepare artificial seeds. This research started in the early 1980s, and after more than 30 years of research, now has some successful experiences.

In fact, artificial seeds are single somatic embryoids wrapped artificially, and their structure is similar to those of natural seeds. As artificial seeds, they should first be well-developed somatic embryoids, that is to say, the embryoids have the potential to develop into complete plants. Somatic embryoids can be obtained from tissue culture or cell culture. It is not enough to have embryos alone, and the germination of embryos needs nutrition, so they also need artificial endosperms that supply the nutrition for embryo germination. In the manufacturing of artificial endosperms, scientists used their thinking to add pest control substances and plant hormones to the artificial endosperms to ensure that future seedlings can grow faster and healthier. In addition, it is necessary to wrap the somatic embryoids and the artificial endosperm, which form the artificial seed coats. It is generally made of polymer materials, which can protect the water in somatic embryoids and artificial endosperms from being lost, and can also avoid damage from external physical factors. Artificial seeds can be created by assembling the three parts of somatic embryoids, artificial endosperms, and artificial seed coats by artificial means.

Artificial seeds and natural seeds are similar in morphology, function, etc. In essence, artificial seeds are homologous to test tube seedlings used for rapid propagation and are products of asexual reproduction. This determines that the artificial seeds have many advantages.

The advantages of artificial seeds are as follows: First, the somatic embryoids produced by plant tissue culture have the advantages of large quantity, fast propagation speed, and complete structure, so it is possible to establish a set of efficient and rapid breeding methods for those famous, special, and excellent plants. Second, somatic embryoids are produced by asexual reproduction. Once excellent genetic traits are obtained, heterosis can be maintained, so that some excellent hybrid seeds can save the trouble of seed production generation by generation, and can be propagated in large quantities and used for a long time. Third, for some elite crops that cannot be promoted and utilized by normal sexual reproduction, such as triploid plants, polyploid plants, and aneuploid plants, artificial seeds can be used to achieve a large amount of reproduction and promotion in a short period of time. Fourth, the few rare and precious excellent varieties obtained through genetic engineering technology and cell fusion technology can also be rapidly and massively propagated in a short time by artificial seed technology. Fifth, in the preparation of artificial seeds, certain nutrients, pesticides, hormones, herbicides, and beneficial microorganisms can be added to promote the growth and development of plants. In addition, since artificial seeds are of single-cell origin and genetically stable, multiple levels of artificial modifications can be carried out during the preparation process, such as introduction of foreign genes, so that the developed plants have new excellent quality. Chemicals such as pesticides and insecticides can be added to the outer layer of the embryoid to make the developed plants have disease resistance.

Artificial seeds can be used for rapid propagation, increasing seed germination rate, and simplifying the complex breeding process of triploid sterile crops such as seedless watermelon. At present, the development of artificial seeds has been fruitful: Artificial seeds such as celery, lettuce, carrots, and broccoli have been successfully developed, and artificial seeds such as celery have been applied to production, and good economic benefits have been achieved; artificial seeds such as rice, rubber, and celery have also been successfully developed in China, and artificial seeds can be germinated and grown into seedlings in the soil, thus opening up a broad prospect for the promotion of the field. In addition, in the production of precious flowers and artificial afforestation, the advantages of artificial seeds are also very obvious.

It is precisely because of the advantages of artificial seeds in simplifying rapid propagation techniques, reducing costs, and facilitating storage, transportation, and mechanized sowing that it is generally considered to be an ideal rapid propagation technique superior to test tube seedlings.

The use of artificial seeds can save a lot of grains. Statistics show that China's annual seed consumption can reach 15 billion kilograms, and these can serve nearly 100 million people's rations for a year. The artificial seeds can produce millions of seeds from the tender bud of a plant, which can save a lot of grains and has a good application prospect.

6. Polyploidy Breeding

There are 23 pairs of chromosomes in human somatic cells, of which 22 pairs are autosomes and one pair is sex chromosomes. Sex chromosomes determine the sex of human beings. Each pair of autosomes consists of two identical chromosomes, and there are also two sex chromosomes, females have two X chromosomes, and males have an X and a Y chromosome. Therefore, human somatic cells are diploid. The number of chromosomes in sperms or eggs is only half that of normal somatic cells, so they are haploid. However, the zygote formed after fertilization becomes a diploid due to the combination of the chromosomes of the sperm and egg. This diploid fertilized egg develops into an embryo. This guarantees that the number of chromosomes in human somatic cells will always be 23 pairs; otherwise, humans will suffer from hereditary diseases.

Of course, there are a few exceptions in nature, such as the bees in insects, whose drones are developed from haploid egg cells, and the drones have normal living ability. In general, haploid individuals are thinner and weaker than diploid parents, have poor living ability, and cannot have offspring. Polyploids are different; organisms of triploid or above are called polyploids, but they are rare in animals and more common in plants. Many plants can form new species by doubling the chromosomes.

The characters of polyploid plants are often different from those of the original diploid plants. In general, the stomata, flowers, fruits, and seeds of tetraploid are larger than those of diploid; the mesophyll is thicker, the stem is thicker, and the metabolites also have obvious changes. For example, the carotenoid content of tetraploid yellow corn is 43% higher than that of the original diploid, the vitamin C content of tetraploid tomato is about twice as high as that of normal diploid, and the sugar content of triploid sugar beet is increased by 14.9% compared with that of diploid sugar beet. These characteristics are exactly what humans need, so the polyploidization of plants is also an important way to cultivate elite crop varieties.

At present, there are two main ways for polyploid breeding. One is through the fertilization of male and female germ cells with no chromosome reduction in the original species or hybrids, and the other is through doubling the chromosome number of the zygotes formed by the fertilization of male and female germ cells of the original species or hybrid, or doubling the chromosome number of somatic cells of the original species or hybrid.

But, what factors can double the number of chromosomes? Scientists have found that there are three main aspects: first, biological factors, such as grafting, distant pollen treatment, and abnormal fertilization; second, physical factors, such as sudden temperature change and radiation exposure; and third, chemical factors, such as chemical treatment. Colchicine is an effective chromosome-doubling agent. Its molecular formula is $C_{22}H_{25}O_6N$, which is a light yellow powder or needle crystal. It can be dissolved in water and alcohol, and belongs to highly toxic substances. The mechanism where colchicine induces the formation of polyploids is to prevent the formation of spindle apparatus during cell division. Spindle apparatus is a device that distributes chromosomes equally into two daughter cells. Once destroyed, the chromosomes that doubled after replication remain in one cell.

Polyploid has many advantages, but it is not a case where the more the multiple, the better the polyploid. Scientists have found that plants above the pentaploid lose their giant traits and show significant signs of decline. The polyploid breeding mainly uses tetraploid and triploid. However, triploid plants cannot give birth to offspring, and it is for this reason that the fruits of some triploid plants are seedless. Musa nana Lour is a typically naturally occurring triploid, so there are no seeds in its fruits. Seedless watermelon is also triploid.

How is seedless watermelon cultivated? We know that the watermelons that we generally eat are diploid. After treating this diploid watermelon with 0.2–0.4% colchicine solution, the tetraploid watermelon with doubled chromosomes is obtained. Then, the common diploid watermelon is used as the male parent and the tetraploid watermelon is used as the female parent to obtain the seeds of triploid seedless watermelon. By planting the triploid seeds in the field like ordinary watermelons, you can grow triploid seedless watermelons. If the variety is properly selected, the grown watermelon is larger and the sugar content is higher.

Polyploid breeding can also be carried out between two plants with a large difference in kinship, for example, the cultivation of octoploid triticale. In fact, wheat and rye are two plants with distant genetic relationships, and the setting rate is low when artificially hybridized. By crossing wheat with rye and trying to double the number of hybrid chromosomes, the octoploid triticale can be obtained, which can fruit. Through planting it in the alpine region of the Yunnan–Guizhou Plateau in the southwestern part of China, the effect of increasing production is very obvious.

The South China Sea Institute of Oceanology of the Chinese Academy of Sciences has successfully cultivated triploid Pinctada martensii, and the treatment group has cultivated more than 50,000 larvae. The induction rate of triploid is more than 90% in the early stage of embryonic development, and the triploid larvae account for about 70%, which has reached the advanced level of similar research in the world. Gonad appearance and histological examination showed that triploid size, body weight, and meat weight were significantly more than diploid; especially the triploid formed by the first polar body increased its shell height, body weight, and meat weight by 13%, 44%, and 58%, respectively.

The scientists of the South China Sea Institute of Oceanology also cultivated pearls for the first time with triploid Pinctada martensii. The preliminary results of the pearl breeding were a 10% reduction in triploid denucleation rate and a 21% increase in qualified round beads, during the one-year beading period. The thickness and weight of the bead layer of the triploid pearl increased by 44% and 55%, respectively. The results laid a scientific foundation for the promotion of pearl culture and artificial culture of oyster, mussel, scallop, abalone, and other marine shellfishes.

Mulberry polyploids, especially triploids, have good economic characters. Radiation treatment could double the chromosome number of the diploid variety of the white mulberry (*Morus alba*), and then when crossed with the diploid mulberry (*Morus multicaulis*), could thereby breed the artificial triploid of the Greater China mulberry. It has the excellent economic characters of typical triploid mulberry with vigorous growth, high yield, good quality, strong stress resistance, cuttage propagation, and easy survival. The 3-year-old and medium density cultivated Greater China mulberry has a leaf yield of more than 3,000 kilograms per mu. Compared with the existing diploid varieties, the yield is increased by more than 30%, and the quality of silkworm cocoons raised by mulberry leaves is also improved.

In 1997, the Institute of Sugar Beet of the Chinese Academy of Agricultural Sciences successfully cultivated the "Sugar-beet-research-single-grain No. 2" polyploid beet hybrid with a sugar content of 15.4% and a sugar yield of 5,695 kilograms per hectare. It has been named and promoted by the Heilongjiang Provincial Crop Variety Approval Committee.

In 2000, the world's first allotetraploid crucian-carp was born in Hunan province, China. The College of Life Sciences of Hunan Normal University and the partner Donghu Fishery, Xiangyin County, Hunan Province, have successfully bred the world's first tetraploid fish population with stable genetic characteristics and natural reproduction by using a combination of cell engineering and sexual hybridization. The tetraploid fish was crossed with the diploid fish, and they successfully bred the sterile triploid crucian (*Xiangyun crucian*) and triploid carp (*Xiangyun carp*). Because this fish does not breed offspring, triploid fishes are also known as environmentally friendly fishes. It has the characteristics of fast growth, good meat quality, high edible rate, strong disease resistance, infertility, and so on. It has been widely promoted in more than 20 provinces and cities in China. A large industrial scale has been formed and significant economic benefits have been achieved. In October of the same year, the appraisal committee headed by academician Zuoyan Zhu who is a well-known genetic breeding expert in China, academician Haoran Lin who is an expert in aquatic biology, and academician Zhonghe Zhai who is an expert in cell biology believed that "this achievement made by Chinese scientists marks a creative breakthrough in the theory and application of polyploid fish breeding, ranking at the international leading level".

In 2007, Shaojun Liu and others from the College of Life Sciences of Hunan Normal University reported that the hybridization of female diploid red crucian (*Carassius auratus*) with male diploid *Megalobrama amblycephala* and their offspring obtained triploid, tetraploid, and pentaploid hybrid fishes. In contrast, the offspring obtained by crossing male diploid Carassius auratus with female diploid *Megalobrama amblycephala* did not survive. The obtained tetraploid hybrid fish can naturally reproduce, and the triploid hybrid fish becomes a new economic fish species.

In 2013, Na Yang and others from the College of Agronomy of Shanxi Agricultural University used the colchicine agarose gel smear method to treat the stem apex growth point of *Gossypium asiatica* seedlings, and obtained *Gossypium asiatica* autotetraploid plants. In 2015, Hong Kong and others from the School of Life Sciences of Langfang Normal College used different concentrations of colchicine solution to treat African Impatiens cuttings, and obtained autotetraploid plants that were significantly different in morphology and cytology from diploid. In 2016, Zhenhua Yang of the Branch of Biotechnology of Yangling Vocational and Technical College used colchicine to treat the germinated seeds of *Glycyrrhiza uralensis* for polyploid breeding.

Although the technique of polyploid breeding appeared in the early 20th century, it is still widely used now. The use of polyploid breeding has undoubtedly enriched the means of plant breeding, and countries around the world have used this method to create many new varieties. In addition to the new polyploid varieties mentioned above, other new triploid varieties include tall triploid *Populus davidiana*, and triploid crabs with body weight more than three times that of common crab. The new tetraploid varieties include tetraploid rubber grass with high rubber content, tetraploid grape, and tetraploid turnip for feed.

The potential of polyploid breeding is still great, and more and more new polyploid varieties will appear over time.

Chapter 6

Ascendant Cell Transplantation Therapy

1. Stem Cells, the Sources of Life

The "stem" in stem cells means "trunk" and "origin". Branches, leaves, flowers, fruits, and so on are all developed from the trunk, and that is to say, the trunk is their origin. Stem cells are the sources of all the other cells (such as muscle cells, nerve cells, and fat cells). From this perspective, stem cells are well-deserved "sources of life."

How were stem cells discovered? This can be traced back to 1867, when Julius Friedrich Cohnheim, a German experimental pathologist, discovered stem cells while studying wound inflammation. He injected animals with the insoluble dye aniline intravenously. As a result, cells containing the dye, including inflammatory cells and fibroblasts related to fiber synthesis, were found in the distal part of the injured animals. Therefore, he inferred that there were stem cells of non-hematopoietic function in the bone marrow. The concept of bone marrow stem cells was first proposed by Julius Friedrich Cohnheim.

In 1974, Alexander Friedenstein and his colleagues first isolated this kind of stem cell from bone marrow, and confirmed that they were different from most hematopoietic cells from bone marrow and can quickly adhere to the culture vessel *in vitro* and produce fibroblast-like clones.

155

They grow in whirlpool shapes *in vitro* and have the ability of self-replication and renewal. Friedenstein and his colleagues also confirmed that after inoculation with bone marrow cell suspension, each stem cell can form a different clone, and there is a linear relationship between the number of stem cell proliferations and the number of colonies. Each stem cell is a colony-forming unit fibroblast (CFU-F), and studied by chromosomal markers, ^{3}H thymidine labeling, delayed photography, and Poisson distribution statistics. Friedenstein encouraged other scientists and doctors to use stem cell transplantation to treat some major diseases.

In 1991, Arnold Caplan officially named this kind of bone marrow cell as "mesenchymal stem cells" (MSCs). He believed that these bone marrow-derived mesenchymal stem cells have the potential to differentiate into bone, cartilage, muscle, bone marrow stroma, tendon/ligament, fat, and other connective tissues. In 2005, the International Society of Cell Therapy announced that the acronym "MSCs" is multipotent mesenchymal stromal cells. Therefore, bone marrow mesenchymal stem cells are sometimes referred to as "bone marrow stromal cells."

But, the real research on stem cells began in the 1960s. In 1963, Canadian scientists McCulloch and Till first demonstrated the existence of stem cells in the blood and found that hematopoietic stem cells can differentiate into hundreds of different types of human tissue cells. In 1981, Kaufman and Martin isolated embryonic stem cells from the inner cell mass of mouse blastocyst, and established suitable culture conditions for embryonic stem cells *in vitro*, and cultivated them into stem cell lines. In the 21st century, the research and the application of stem cells have been given great attention and have become the target of reports by major media. What are stem cells? A more rigorous definition can be given: Stem cells are a group of primitive cells with self-renewal and multi-directional differentiation potential. This definition has two meanings: One is that stem cells have the ability of self-renewal and replication, or that they can self-propagate to produce new stem cells; the other is that stem cells can differentiate into other cells and then form tissues, organs, and even individuals. For example, stem cells can differentiate into cardiomyocytes with beating function or differentiate into islet cells with insulin-secreting function. So, stem cells are young cells that have not become mature yet.

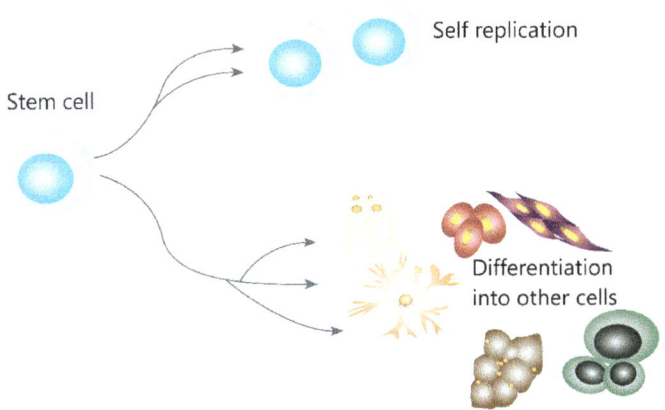

Stem cells are cells that can self-replicate and differentiate into other cells

Stem cells are a large family. There are many different kinds of stem cells, and the methods of classification are varied. Two main classification methods are described below.

According to the differentiation potential, stem cells can be divided into four types: Totipotent stem cells, sub-totipotent stem cells, multipotent stem cells, and unipotent stem cells. They can also be divided into three types, namely, totipotent stem cells, multipotent stem cells, and unipotent stem cells. But, it may be more scientific to divide them into four types.

The first is totipotent stem cells, which are the most powerful stem cells and the only stem cells that can develop into complete individuals. They can proliferate and differentiate into the cells of all tissues and organs and form complete individuals. Where can we find totipotent stem cells? The cells from the fertilized egg to the 32-cell cleavage stage embryos are totipotent stem cells, which are also the main source of totipotent stem cells. In addition, germ cells are also totipotent stem cells, such as the eggs of a queen bee. There are many worker bees, a few male bees, and only one queen bee in a bee colony, among which the queen bee and worker bees are female. Although the queen bee and worker bees are all females, the queen bee has fertility, and the worker bee does not. The reason for this difference is that they eat different foods when they are young. The female larvae that develop into queens will continue to eat royal jelly until the development is complete, while the female larvae that develop into worker bees do not. The eggs laid by the queen bee develop into queen bees or worker bees if fertilized, and male bees when unfertilized. The egg cells of the queen bee are totipotent stem cells, because they can differentiate into the cells of all tissues and organs, and have the ability to develop into the whole individuals. Another example is Dolly, a cloned sheep. It was developed by transplanting somatic cell nucleus into an enucleated egg cell. This "egg cell" has the ability to proliferate and differentiate into all tissues and organs and develop into an individual, so it is a totipotent stem cell.

The second is sub-totipotent stem cells, or pluripotent stem cells. Their ability to develop and differentiate is only second to totipotent stem cells. They can participate in the multisystem differentiation of three germ layers into mature tissue cells, such as skin tissue, nerve tissue, lung tissue, liver tissue, hematopoietic cells, muscle cells, and osteoblasts, so as to achieve regeneration and functional reconstruction of various tissue lesions or injuries. Embryonic stem cells are a kind of stem cells isolated from early-stage embryos. They can develop into any kind of cells of ectoderm, mesoderm, and endoderm, but they cannot develop into a complete individual only by themselves, so they are called sub-totipotent stem cells. Induced pluripotent stem cells, also called iPS cells, are induced by cloning four transcription factor genes of Oct3/4, Sox2, c-Myc, and Klf4 into viral vectors and then transferring them into mouse fibroblasts. Such cells are similar to embryonic stem cells in terms of morphology, gene and protein expression, epigenetic modification status, cell-doubling ability, embryoid body and teratoma formation ability, differentiation ability, etc., so they should also be sub-totipotent stem cells.

The third is multipotent stem cells. Multipotent stem cells have the potential to differentiate into a variety of tissue cells, but have lost the ability to develop into complete individuals. They cannot develop into tissues and organs of all three germ layers, and the developmental potential of them is further restricted. Bone marrow hematopoietic stem cells are a typical example. They can differentiate at least twelve kinds of blood cells, but they cannot differentiate into cells other than the hematopoietic system.

The fourth is unipotent stem cells. Unipotent stem cells are also called specialized potential stem cells. Such stem cells can only differentiate into one type or two closely related types of cells, such as the stem cells of the basal layer of the epithelial tissue and the myoblasts in muscle. Unipotent stem cells are the stem cells with the lowest developmental differentiation level and have poor self-renewal ability.

In theory, totipotent stem cells can be differentiated into pluripotent stem cells (sub-totipotent stem cells), multipotent stem cells, and unipotent stem cells. Pluripotent stem cells can differentiate into multipotent stem cells and unipotent stem cells, and multipotent stem cells can differentiate into unipotent stem cells. Therefore, in terms of development and differentiation potential, the level of totipotent stem cells is the highest, and that of unipotent stem cells is the lowest.

According to different sources, stem cells are classified into embryonic stem cells and adult stem cells.

The embryonic stem cells is a type of cell isolated from the early embryo (before gastrula stage) or primordial gonad, which has the characteristics of indefinite proliferation, self-renewal, and multi-directional differentiation *in vitro*. Whether *in vitro* or *in vivo*, embryonic stem cells can be induced to differentiate into almost all cell types in the organism. But, embryonic stem cell research has always been a controversial field. The proponents believe that this research will help to eradicate many intractable diseases, because the embryonic stem cells can differentiate into many kinds of functional cells, which are considered to be a life-saving charity and a manifestation of scientific progress. The opponents argue that embryonic stem cell research must destroy embryos, and embryos are the life forms of human beings in the womb when they are not yet formed, which has ethical problems. Adult stem cells include mesenchymal stem cells, hematopoietic stem cells, neural stem cells, adipose stem cells, skin stem cells, hair follicle stem cells, and limbal stem cells, which exist in various tissues and organs of an organism, such as umbilical cord blood, bone marrow, and adult organ tissues. Adult stem cells in adult tissues and organs are mostly dormant under normal conditions, but they can exhibit different degrees of regeneration and renewal ability under pathological conditions or induced by external factors.

It is precisely because various stem cells have the ability to develop and differentiate into other cells, tissues, and organs that stem cells are called

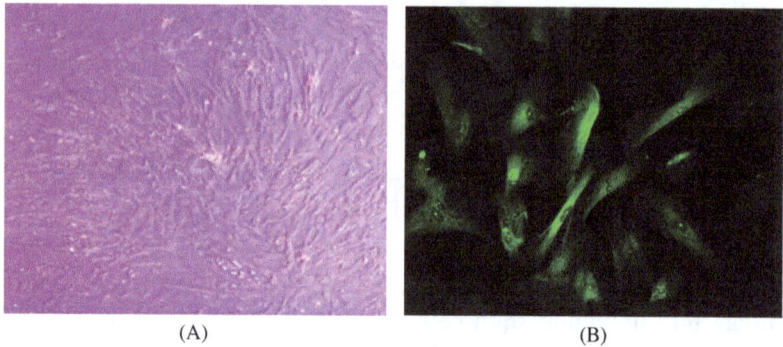

(A)   (B)

Human amniotic mesenchymal stem cells

(A) Observation under low-power microscope; (B) observation under high-power microscope.

universal cells in medicine. They can be used to treat some intractable diseases, repair or reconstruct tissue organs, aid in anti-aging, and enhance beauty.

2. Stem Cell Storage, Bank of Life

Leukemia is commonly known as "blood cancer". The prevalence rate of urban and rural residents in China is 3.82/100,000, ranking sixth in the mortality rate of malignant tumors. Among them, there are more men than women, people in cities are slightly higher than those in rural areas, and the incidence rate and mortality rate in children and adolescents are much higher. It seriously endangers human health. An effective treatment for leukemia is bone marrow transplantation, but bone marrow sources are limited and expensive. So, people have stored their bone marrow in advance for future use. This kind of special institution for storing bone marrows is called a bone marrow bank.

The Bone Marrow Donors Worldwide (BMDW) was established in 1994 with headquarters in Leiden, the Netherlands. It is a voluntary organization, and bone marrow banks of all countries can participate in voluntarily, aiming at eliminating the obstacles of cross-border inquiry, donation, and transplantation, so as to enable the exchange, discussion and common development of bone marrow banks in different countries.

Human stem cell bank
(Saliai Stem Cell Research Institute, Guangdong Province, China)

The largest bone marrow bank is the National Marrow Donor Program (NMDP), which is headquartered in Minneapolis, Minnesota. It was established in 1986 and has more than 7 million volunteers. The donation methods include bone marrow donation and peripheral blood hematopoietic stem cell donation. The annual donation amounts are more than 4,000 cases. The second is the German Bone Marrow Donor Center or Deutsche Knochenmarkspenderdatei (DKMS), with more than 3.6 million volunteers and more than 3,000 donation cases each year. The China Marrow Donor Program (CMDP) is currently the fourth largest bone marrow bank in the world after the United States, Germany, and Brazil. Its predecessor was the "Chinese unrelated bone marrow transplantation donor database" approved to be established by the Ministry of Health in 1992. According to the information released by the website of the China bone marrow bank, as of September 30, 2020, the capacity of the China bone marrow bank was 2,812,573 persons, and a total of 10,286 volunteers donated hematopoietic stem cells to patients. The hematopoietic stem cells are a type of stem cells found in the bone marrow hematopoietic tissue, which can be differentiated into other types of blood cells, such as red blood cells, white blood cells, and platelets.

Liquid nitrogen tanks for storing stem cells

The strange thing is, why is it called the bone marrow bank, but what actually is donated is hematopoietic stem cells?

Yes, because the bone marrow bank is also known as the hematopoietic stem cell donor database bank, and also accepts hematopoietic stem cell donation. It turns out that with the development of medicine, researchers found that the effective ingredients in the treatment of leukemia are mainly hematopoietic stem cells in the bone marrow, although a small number of mesenchymal stem cells also play certain roles. Therefore, hematopoietic stem cells can be directly isolated and purified from bone marrow for clinical treatment of leukemia. The advantages of doing so are as follows: First, it reduces the probability of immune rejection in exogenous bone marrow and improves the success rate of transplantation. This is because the components of bone marrow are complex, and many components in exogenous bone marrow can cause immune rejection after transplantation; second, hematopoietic stem cell transplantation does not need to be matched like bone marrow transplantation, which simplifies clinical procedures, because hematopoietic stem cells are a type of primitive cells with weak immunogenicity; and third, the source of bone marrow is extremely limited, but the source of hematopoietic stem cells is much more abundant, which provides more clinical options. Once a baby is born, hematopoietic stem cells can be extracted from umbilical cord blood or placenta for preservation. The number of hematopoietic stem cells in umbilical cord blood or placenta is large and energetic. It is also a waste, which can be discarded as medical waste if it is not used. Hematopoietic stem cells can also be extracted from adult peripheral blood or bone marrow, but there will be some pain, and the activity of hematopoietic stem cells extracted from the adults is relatively weak. These sources of hematopoietic stem cells, whether autologous or allogeneic, can be used for clinical hematopoietic stem cell transplantation for the treatment of leukemia.

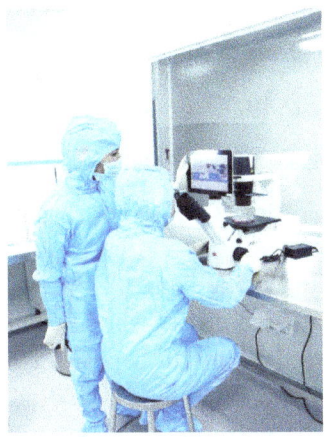

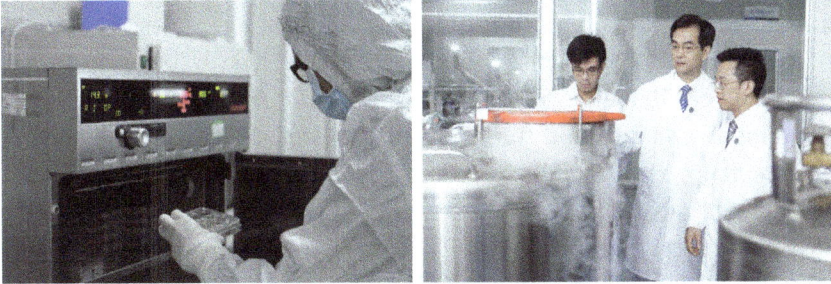

The center for preparation and storage of clinical-grade stem cells

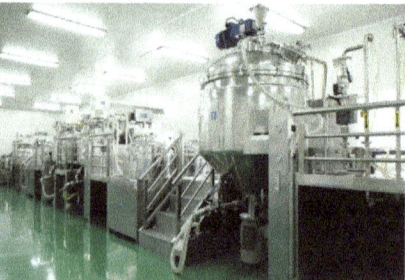

GMP workshops for the preparation of stem cell products

The cost of hematopoietic stem cell transplantation is very high, which is hard to be borne by the general family. However, because of the unpredictability of leukemia, no one knows whether leukemia will occur in the future. It is like going to the bank to deposit money. You can save your own hematopoietic stem cells beforehand and take them out when you need them. For the sake of prevention, many families choose to store the umbilical cord blood stem cells of their children at their birth.

The umbilical cord hematopoietic stem cell bank, referred to as the umbilical cord blood bank, is a special institution to extract and preserve cord blood hematopoietic stem cells, reserve resources for patients who need hematopoietic stem cell transplantation, and provide stem cell matching query.

To date, there are more than 150 umbilical cord blood banks in the world, including 40% in Europe, 30% in North America, 20% in Asia, and 10% in Oceania. The United States is the earliest country to build umbilical cord blood banks. In 1993, Rubinstein established the world's first public umbilical cord blood bank in the New York Blood Center. Currently, there are 32 private umbilical cord blood banks and 31 public umbilical cord blood banks. According to the statistics, the United States stores about 500,000 umbilical cord blood samples per year, accounting for about 2.6% of all the newborn babies. European umbilical cord blood bank construction was also relatively early. Currently, Europe has the largest number of umbilical cord blood banks in the world, with more than 29 private umbilical cord blood banks and 50 public umbilical cord blood banks. Japan's first umbilical cord blood bank was established in 1994. A union of umbilical cord blood banks was established in 1999, which has preserved more than 20,000 umbilical cord blood samples. According to the data of the Bone Marrow Donors Worldwide (BMDW) in April 2003, 32 umbilical cord blood banks in 21 countries have preserved more than 130,000 umbilical cord blood samples. The construction of umbilical cord blood banks in China started in 1998. At present, the approved seven umbilical cord blood banks are distributed in 7 provinces and municipalities, including Beijing, Tianjin, Shanghai, Guangdong, Sichuan, Shandong, and Zhejiang. More provincial and municipal umbilical cord blood banks are under planning and construction.

The institutions responsible for storing stem cells, such as the bone marrow bank and the umbilical cord blood bank, are called stem cell banks, because they are similar to banking businesses, and are also called "life bank". There is interest for saving money in banks, but there is no interest for storing stem cells, and even a custody fee needs to be paid. For the customers, they have to pay for the protection, and the intention is that it can be used in the future, but of course, it may never be used. From this point of view, the businesses of stem cell banks are similar to those of insurance companies. There are many kinds of stem cells and stem cell banks. According to the types and sources of the prepared and stored stem cells, the stem cell banks can be divided into bone marrow bank, umbilical cord blood bank, embryonic stem cell bank, placental stem cell bank, induced pluripotent stem cell bank, mesenchymal stem cell bank, deciduous dental pulp stem cell bank, and general stem cell bank. The stem cell banks can be divided into public stem cell banks and private stem cell banks depending on the supply manner and application objects. The stem cells stored in the public stem cell banks are the stem cells donated by others to meet the needs of patients who did not preserve the autologous stem cells before, while the private stem cell banks store the patients' own stem cells for use by themselves.

The various stem cell banks provide some guarantee for disease treatment and human health protection.

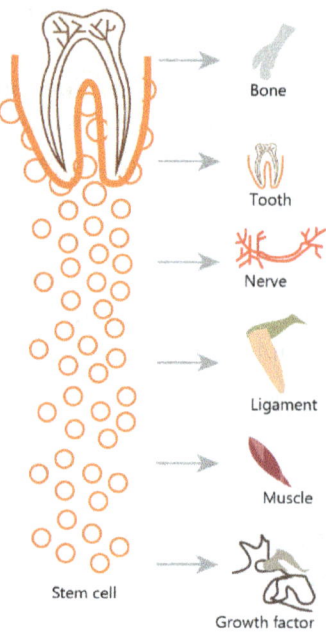

Dental pulp stem cells

3. Stem Cells, Biological Drugs

In the human body, stem cells can be induced to transform into other functional cells, thereby repairing or regenerating the dysfunctional and missing tissues and organs for the purpose of treating diseases. As a drug, stem cells have the advantages of low side effects and good safety, and can achieve therapeutic effects that are difficult to achieve by other treatment methods. Several kinds of new stem cell drugs have been put on the market.

Human embryonic stem cells (hESCs) are very ideal stem cell types for developing stem cell therapy. The American companies of Geron and ACT are the earliest enterprises to develop human embryonic stem cell drugs.

Embryonic stem cell technology is the main way to develop chronic degenerative disease therapies. Chronic degenerative diseases have the characteristics of strong concealment, complex pathogenic factors, and preventability. The company Geron has developed unique methods for the production, maintenance, and scaling up of the undifferentiated human embryonic stem cells. The company Geron has also developed ways to differentiate these human embryonic stem cells into the cells needed for therapy and to refrigerate the differentiated cells for commercial sale.

Geron has developed six kinds of stem cell drugs. These cell drugs are derived from human embryonic stem cells by using different methods. Among them, a drug for the treatment of nervous system diseases such as spinal cord injury has entered phase I clinical trial. In January 2009, the US Food and Drug Administration gave approval to the American company Geron to carry out a clinical trial of human embryonic stem cell therapy for spinal cord injury, which became the first approved human embryonic stem cell therapy in the United States. There are more than 20 kinds of drugs developed by the company Geron in the United States, which can treat non-small cell lung cancer, chronic lymphocytic leukemia, multiple myeloma, thrombocytosis, breast cancer, and other solid tumors.

On March 2, 2010, the American company ACT announced that its embryonic stem cell therapy for the treatment of juvenile blindness has also won the orphan drug status granted by the US food and Drug Administration. An orphan drug is a drug for the treatment of rare diseases. In November 2009, the company ACT submitted to the US Food and drug administration a clinical trial application for the treatment of recessive macular dystrophy by using human embryonic stem cells to reconstruct retinal pigment epithelial cells to achieve the treatment of this disease. This news opened the doors for the industrialization of stem cell therapies, and also brought hope for patients with diseases.

However, due to ethical constraints, many enterprises cannot use human embryonic stem cells to develop stem cell therapies, or the developed human embryonic stem cell therapies have difficulty in obtaining approval. For this reason, the development of human adult stem cell drugs has received increasing attention and achieved remarkable results.

In 2011, human adult stem cells accounted for more than 80% of the whole stem cell market. Compared with the human embryonic stem cells, the human adult stem cells simply operate in the collection process, have a small probability of contamination in the process of cultivation, and do not have ethical problems. Moreover, human adult stem cell therapy does not need genetic manipulation, and the medication is safe.

The biggest advantage of human adult stem cells is that the genome is very stable. The mesenchymal stem cells are rich in sources and can be isolated from bone marrow, adipose tissue, epidermis, blood, and other tissues, and can produce precursor cells of bone, cartilage, fat and blood cells, and fibrous connective tissue, and have broad development prospects. For these reasons, mesenchymal stem cells have become the best choice for the regenerative medicine industry to develop commercial products in the short term.

Allogeneic cell therapy products have been used to treat graft-versus-host disease, bone marrow transplantation, and diabetic ulcers. More importantly, allogeneic mesenchymal stem cell therapy has entered phase II and phase III clinical trials of the therapy of degenerative indications such as rheumatoid arthritis, diabetes, ischemic heart disease, osteoarthritis, and muscle injury.

The American Osiris Therapeutics Incorporation is mainly engaged in the study of obtaining mesenchymal stem cells from adult bone marrow. At present, the company's products have proven to have the ability to repair different types of tissues and provide opportunities for the development of innovative therapies for a variety of diseases such as inflammatory diseases, heart disease, diabetes, and arthritis.

Osiris Therapeutics Incorporation has developed two relatively mature stem cell products, Prochymal and Chondrogen, and has conducted numerous clinical trials. Prochymal is adult mesenchymal stem cell derived from bone marrow that has the effect of controlling inflammation, promoting tissue regeneration, and preventing scar formation. Prochymal has entered or completed phase III clinical trials in the treatment of four diseases, including graft-versus-host disease and Crohn's disease.

The drug can also be used to repair myocardial tissue after heart attack, protect islet cells in patients with type 1 diabetes, and repair lung tissue for patients with lung disease. The efficacy and safety of Prochymal in the treatment of type 1 diabetes has been recognized by the US Food and Drug Administration. On May 4, 2010, the US Food and Drug Administration authorized Prochymal to enter the clinical treatment of type 1 diabetes. Chondrogen is mainly used to treat arthritic disease and the like. The Phase I clinical trial for the treatment of knee arthritis with this drug has been completed, and the recruitment of patients for phase II clinical trial has also been completed.

The American company Stem Cells has developed human neural stem cells. The company's product HuCNS-SC is highly purified human neural stem cells isolated from fetal brain. The preclinical studies have confirmed that these cells can be directly transplanted into the central nervous system and can differentiate into neurons and glial cells. They can survive in the body for up to a year without any tumor formation or any side effects. This product has entered the clinical trial stage in the clinical treatment of two kinds of serious nervous system disorder diseases. In January 2009, the company Stem Cells completed phase I clinical trial of the product for the treatment of neuronal ceroid lipofuscinosis. In November 2009, the company launched a phase I clinical trial of using HuCNS-SC in the treatment of familial centrolobar sclerosis, which affects the myelin disorders of young children. The company is also developing stem cell therapies for the treatment of Alzheimer's disease and age-related macular degeneration.

Due to the large market of cardiovascular diseases, the development of stem cell transplantation in the treatment of cardiovascular diseases is the focus of research and development of biotechnological companies. In the near future, there may be multiple mesenchymal stem cell products that will be approved. The company Bioheart in the United States has developed two cell products that can repair heart injury. Among them, MyoCell is a muscle stem cell that can improve heart function in patients after severe heart injury for several months or years. The Australian company Mesoblast's bone marrow mesenchymal stem cell product, a kind of umbilical cord blood hematopoietic stem cell therapy product, has entered

Phase III clinical trial for the treatment of myocardial infarction. Intramyocardial injection of autologous hematopoietic stem cells by American company Baxter has undergone a phase III clinical trial to improve myocardial blood flow and relieve angina pectoris in patients with refractory chronic myocardial ischemia. Current research indicates that the product repairs heart tissue, increases blood flow, reduces the risk of angina attacks, and enables patients to exercise properly.

The company ACT in the United States currently focuses on three areas, namely, the reconstruction of retinal pigment epithelial cells, hemangioblasts, and myoblasts. In the field of hemangioblast research, ACT is currently conducting preclinical trials to treat cardiovascular disease, stroke, and cancer. At present, the company has successfully used human embryonic stem cells to obtain vascular endothelial cells and successfully used them in vascular repair. Myoblast research was initiated by ACT in 2007. In September 2007, the company ACT acquired the company Mytongen and took over the development of heart failure therapy. The project mainly uses stem cells to obtain myoblasts to repair heart damage caused by heart failure.

In addition to the abovementioned approved stem cell drugs and products, stem cell drugs and products in the middle and late clinical research period also include the StemEx of the company Gamida Cell in Israeli, which is an allogeneic stem cell product for the treatment of leukemia and lymphoma. In the future, leukemia patients will be able to use their own stem cells for bone marrow transplantation, instead of having to purchase stem cells at a huge cost from volunteers who have successfully matched them. Therefore, it is also regarded as a ready-to-use treatment.

In recent years, many pharmaceutical enterprises and scientific research institutes in China have tried to develop stem cell drugs. According to the information provided by the website of the China Food and Drug Administration, some drugs have been clinically studied. From March to April 2014, the Institute of Basic Medical Sciences of the Chinese Academy of Medical Sciences carried out a study on bone marrow mesenchymal stem cells in the prevention of acute graft-versus-host disease, and completed the non-randomized and randomized phase II clinical trials, with the indications for malignant hematologic diseases and

Some stem cell drugs being developed

Product	Source	Product Description	Indication	Development Stage	Producer
GRNOPC1		Oligodendrocyte progenitor cells	Spinal cord injury	Phase I	Geron Corporation
GRNCM1		Cardiomyocytes	Heart disease	Preclinical	
GRNIC1		Islet cells	Type 1 diabetes	Research	
GRNCHND1		Chondrocytes	Osteoarthritis	Research	
		Stem cells	ADME drug screening	Research	
GRNVAC1/ GRNVAC2		Mature dendritic cell	Tumor immunotherapy	Research	
		Immature dendritic cells	Immune Rejection	Research	
Osteoblasts		Osteoblast	Osteoporosis	Research	
Prochymal	Mesenchymal stem cells	Bone marrow mesenchymal stem cells	Hormone resistant acute graft-versus-host disease	Phase III	Osiris Incorporation
			First-line treatment of acute graft-versus-host disease	Phase III	
			Refractory Crohn's disease	Phase III	
			Type 1 diabetes mellitus	Phase III	
			Acute myocardial infarction	Phase II	
			lung disease	Phase II	
			Acute radiation syndrome	Phase III (Animal Rule)	
Chondrogen	Mesenchymal stem cells		Osteoarthritis and cartilage protection	Phase II	
Osteocel-XC	Mesenchymal stem cells		Local bone regeneration	Preclinical	

(Continued)

(Continued)

Product	Source	Product Description	Indication	Development Stage	Producer
Provacel	Mesenchymal stem cells		Myocardial injury	Phase I	
ReN001	Central nervous stem cell		Stroke	Phase I	ReNeuron Incorporation
ReN009			Peripheral arterial disease	Preclinical	
ReN003			Retinal blinding disease	Preclinical	
HuCNS-SC	Central nervous stem cell		Neuronal ceroid lipofuscinosis	Phase I	Stem Cells Incorporation
			Familial middle lobe sclerosis	Phase I	
MyoCell®	Muscle stem cell		Class II / Class III heart failure	Phase II / Phase III	Bioheart Incorporation
Osteocel® Plus	Mesenchymal stem cells		Musculoskeletal defects	Phase III	NuVasive Incorporation
MA09-hRPE	Human embryonic stem cell		Recessive macular dystrophy	US Food and Drug Administration approved clinical trials	Advanced Cell Technology Incorporation
Myoblas	Human embryonic stem cell		Heart failure	Phase III	
AMR-001	Hematopoietic stem cell		ST-segment elevation myocardial infarction	Phase II	NeoStem Incorporation
Ixmyelocel-T	Autologous bone marrow cells		Severe limb ischemia and dilated cardiomyopathy	Phase III	Aastrom Incorporation
C-Cure	Mesenchymal stem cells		Heart failure	Phase II / Phase III	Cardio 3Biosciences Company

graft-versus-host disease. The phase I clinical trial of "Mesenchymal Stem Cell Myocardial Infarction Injection" was completed in 2011 by Hebei Better Cell Biotechnology Co., Ltd., and the safety and efficacy of stem cell therapy for myocardial infarction were preliminarily evaluated. The indications included cardiac insufficiency during the recovery period of acute myocardial infarction. However, the China Food and Drug Administration has been cautious about the clinical trials of stem cell drugs declared by domestic pharmaceutical companies, and the process of approval of relevant research is still relatively tortuous. Despite this, there will still be a high probability that new stem cell drugs will be available in the next few years.

Since the research of stem cell drugs was launched, a variety of new stem cell drugs have been marketed, and many new drugs are in the phase III clinical research stage.

Some of the cell drugs that have been approved

Country or Region	Date	Trade name (Company)	Source of Cells	Indications
European Medicines Agency (EMA)	2009.10	ChondroCelect (TiGenix, Belgium)	Autologous chondrocyte	Knee articular cartilage defect
US Food and Drug Administration (FDA)	2009.12	Prochymal (Osiris, USA)	Human allogeneic bone marrow derived stem cells	Graft versus host disease (GVHD) and Crohn's disease
Australian Therapeutic Goods Administration (TGA)	2010.07	MPC (Mesoblast, Australia)	Autologous mesenchymal precursor cells	Bone repair
Korea Food and Drug Administration (KFDA)	2011.07	Hearticellgram-AMI (FCB-Pharmicell, Korea)	Autologous bone marrow mesenchymal stem cells	Acute myocardial infarction (AMI)
US Food and Drug Administration (FDA)	2011.11	Hemacord (New York Blood Center, USA)	Allogeneic umbilical cord blood hematopoietic progenitor cells	Hereditary or acquired hematopoietic diseases

(Continued)

(Continued)

Country or Region	Date	Trade name (Company)	Source of Cells	Indications
Korea Food and Drug Administration (KFDA)	2012.01	Cartistem (Medipost, Korea)	Umbilical cord blood derived mesenchymal stem cells	Degenerative arthritis and knee articular cartilage injury
Korea Food and Drug Administration (KFDA)	2012.01	Cuepistem (Anterogen, Korea)	Autologous adipose derived mesenchymal stem cells	Complex Crohn's disease complicated with anal fistula
Health Canada (HC)	2012.05	Prochymal (Osiris, USA)	Allogeneic bone marrow stem cells	Children acute graft versus host disease (GVHD)
European Medicines Agency (EMA)	2015.02	Holoclar (Chiesi Farmaceutici, Italy)	Human corneal Epithelial cells (containing stem cells)	Moderate to severe Limbal stem cell deficiency (LSCD) caused by physical or chemical burning in adult patients
European Medicines Agency (EMA)	2015.06	Stempeusel (Stempeutics Institute, India)	Allogeneic bone marrow stem cells	Thromboarteritis obliterans
Japan's Ministry of Health, Labour, and Welfare (MHLW)	2015.09	Temcell (JCR Pharmaceuticals Co., Japan)	Allogeneic bone marrow stem cells	Acute graft-versus-host disease (GVHD), one of the serious complications after hematopoietic stem cell transplantation
US Food and Drug Administration (FDA)	2016.12	Maci (Vericel, USA)	Autologous chondrocytes (A tissue engineering product cultured on collagen membrane of pig)	Knee articular cartilage injury
US Food and Drug Administration (FDA)	2017.08	Kymriah™ (tisagenlecleucel; Novartis, Switzerland)	Chimeric antigen receptor T cells (CAR-T)	3–25 years of patients with acute lymphoblastic leukemia

(Continued)

Country or Region	Date	Trade name (Company)	Source of Cells	Indications
US Food and Drug Administration (FDA)	2017.10	Yescarta (axicaba-geneciloleucel; Kite, USA)	Chimeric antigen receptor T cells (CAR-T)	Adult patients with relapsed or refractory large B-cell lymphoma
European Commission (EC)	2018.03	Alofiselt (TiGenix, Belgium; Takeda, Japan)	Allogeneic adipose derived mesenchymal stem cells	Adult patients with non-active / mildly active luminal Crohn's disease complicated with complex perianal fistula
Drug Controller General of India (DCGI)	2020.08	Stempeucel (Cipla, India; Stempeutics Research, India)	Adult allogeneic bone marrow mesenchymal stem cells	Critical limb ischemia (CLI) due to buerger's disease and atherosclerotic peripheral arterial disease

In the field of stem cell new drug research, although there is still a gap between China and a few developed countries, Chinese scientists have closely followed the pace of international advanced technology and are becoming leaders in certain fields. A number of new stem cell drugs will be put on the market in the next few years.

It is estimated that in the future, people can use their own or other people's stem cells and stem cell-derived tissues and organs to replace diseased or aging tissues and organs, and they can be widely used to treat a variety of intractable diseases that are difficult to treat with current medical methods, such as leukemia, lymphoma, pernicious anemia, Alzheimer's disease, Parkinson's disease, diabetes, cirrhosis, stroke, and spinal cord injury, which are serious diseases that are currently incurable.

4. Stem Cells, Clinical Applications

The reason why stem cells have attracted widespread attention all over the world is their huge potential for clinical applications. Stem cells have a good treatment effect on the difficult and complicated diseases that are ineffective or have poor efficacy in some traditional medicines (including western medicine and traditional Chinese medicine). Moreover, stem cell transplantation has a wide range of clinical applications, including blood system diseases, nervous system diseases, immune system diseases, cardiovascular diseases, digestive diseases, lower limb ischemia, plastic surgery, anti-aging, and other clinical research fields, which are incomparable with any other drugs.

Leukemia also known as blood cancer or blood tumor belongs to the blood system diseases. Any disease that originates in the hematopoietic system or affects the hematopoietic system with abnormal blood changing, characterized by anemia, bleeding, and fever, is called blood system disease, referred to as blood disease. In order to make patients recover the hematopoietic function as soon as possible and save their precious lives, it is necessary to transplant hematopoietic stem cells. Transplantation of hematopoietic stem cells (including umbilical cord blood hematopoietic stem cells) is a safe and effective method for treating diseases of the blood system, but this technology is difficult and risky [15], and it has higher requirements for the service ability of medical institutions and the technical levels of personnel.

Unpublished data from the European Acute Leukemia Working Party (ALWP) suggest that the 5-year leukemia-free survival rate, overall survival rate, recurrence rate, and transplantation-related mortality rate of 2100 patients with complete remission of acute myeloid leukemia who received autologous peripheral blood stem cell transplantation from 1996 to 2001 were 43%, 76%, 53%, and 9%, respectively. However, the efficacy of chemotherapy as a consolidation and intensification regimen was much lower than the above data; even the disease-free survival rate of low-risk patients at complete remission stage after chemotherapy was lower than 40% to 60%. A phase III, randomized, multicenter clinical trial was conducted by Dr. Anasetti and other researchers from the Moffitt Cancer Center in the United States to compare the 2-year survival rates of peripheral blood stem cell transplantation and bone marrow transplantation from unrelated donors. A total of 551 leukemia patients were randomly divided into groups for the treatment of peripheral blood stem cell transplantation or bone marrow transplantation according to the ratio of 1:1. The results showed that the 2-year overall survival rate was 51% in the peripheral blood stem cell transplantation group and 46% in the bone marrow transplantation group. The researchers found that the total transplant failure rates in the peripheral blood stem cell transplantation group and bone marrow transplantation group were 3% and 9%, respectively. The research of the National Stem Cell Engineering Technology Research Center in China has shown that low-intensity pretreatment of haploidentical hematopoietic stem cells combined with umbilical cord mesenchymal stem cell transplantation in the treatment of severe aplastic anemia can significantly improve the efficacy and reduce the occurrence of complications. In recent years, the range of indications for hematopoietic stem cell transplantation has been expanded, and many refractory diseases have made effective attempts, and achieved good results.

Stem cells can differentiate into neurons and glial cells, regenerate damaged axons, various extracellular matrices, and myelin, maintain the integrity of nerve fiber function, and thus can be used to treat neurological diseases. It is reported that the nervous system diseases treated with stem cells include amyotrophic lateral sclerosis [16], multiple sclerosis, spinal cord injury, Parkinson's disease, schizophrenia, cerebral infarction sequela, cerebellar atrophy, cerebral palsy, stroke sequelae, intracranial hematoma sequelae, hemiplegia, Alzheimer's disease, ataxia, and myasthenia gravis.

Jiang and other researchers performed mesenchymal stem cell transplantation on 20 patients with spinal cord injury, and the results showed that the sensory, motor, and autonomic functions of the patients were significantly improved. The clinical trial of stroke patients reported by Honmou and others found that autologous bone marrow mesenchymal stem cells can reduce the lesion volume by about 20% one week after the transplantation. The patients in the transplantation group have different degrees of the recovery of nerve function, which confirmed the efficacy and safety of bone marrow mesenchymal stem cell transplantation. In a hospital in China, 127 patients with spinal cord injury and 25 patients with ischemic brain injury were treated with autologous mesenchymal stem cell transplantation. It was found that mesenchymal stem cell transplantation was safe and effective, the symptoms of postoperative follow-up were improved, and the motor and sensory functions were restored in varying degrees. The effect of stem cell transplantation within one month after injury was the most obvious. The longer the time after injury, the less significant the curative effect was, but there were no adverse reactions. At the same time, other researchers on stem cell transplantation for other nervous system diseases have also proved the safety and effectiveness of stem cell therapy.

Traditional methods are not ideal for the treatment of immune system diseases, but some stem cell transplantation therapy has better curative effects. The main autoimmune diseases that can be treated by stem cell transplantation are systemic lupus erythematosus, rheumatoid arthritis, systemic sclerosis, diabetes, etc.

Clinical studies have shown that the transplanted mesenchymal stem cells can significantly improve the symptoms of patients with rheumatoid arthritis. A new study by Wang and others explored the efficacy and safety of mesenchymal stem cell therapy in patients with rheumatoid arthritis. A total of 172 patients with rheumatoid arthritis who did not respond well to traditional treatment were randomly divided into two groups: the traditional anti-rheumatic drug group and the umbilical cord mesenchymal stem cell group. The results showed that there was no obvious adverse reaction during and after transplantation of stem cells, the percentage of $CD4^+$ $CD25^+$ $Foxp3^+$ regulatory T cells in peripheral blood increased, and the disease was significantly relieved. Liang and other researchers used allogeneic mesenchymal stem cells to treat 15 patients with refractory systemic lupus erythematosus (SLE) to study the efficacy and safety. The results showed that after treatment with mesenchymal stem cells, the clinical symptoms of all patients were significantly improved, the disease activity index and 24-hour urinary protein were significantly reduced, and no treatment-related adverse reactions were found during the treatment. At present, a number of research institutions in Europe and North America have started the stem cell treatment of systemic sclerosis phase II and phase III clinical trials. For example, the latest clinical trial ASSIST study is a North American Phase II clinical trial to evaluate the safety and efficacy of autologous stem cell transplantation in the treatment of systemic sclerosis. The results showed that although the risk of transplant-related death was higher, its long-term survival benefit was significant.

In recent years, Dianliang Wang and other researchers have made important progress in the treatment of type 1 diabetes with human amniotic mesenchymal stem cells. Through the transplantation therapy experiment of tail vein, hepatic portal vein, and the beneath renal capsule, it was found that human amniotic mesenchymal stem cells can significantly reduce the blood sugar level of diabetic rats and alleviate the symptoms of diabetes. Through the localization and tracking *in vivo*, it was found that stem cells could home to the injured pancreas and repair the islets. All diabetic rats were treated effectively. At present, further clinical studies are under way.

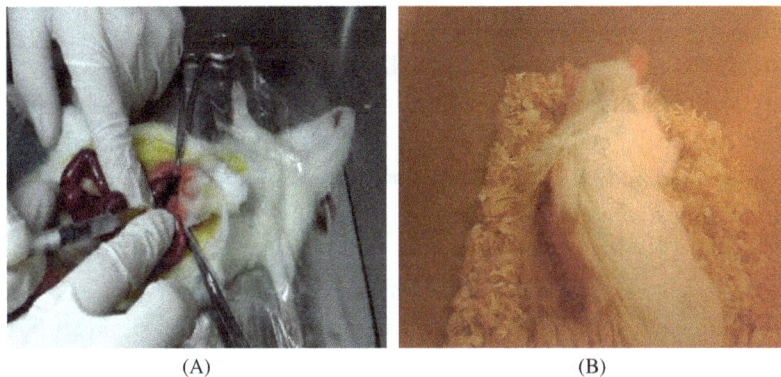

(A) (B)

Stem cells were injected into the rat hepatic portal vein by Dianliang Wang and other researchers to study the treatment of diabetes

(A) The stem cells were transplanted through injection after surgical operation; (B) the Rat after operation.

Some stem cells are used to treat cardiovascular diseases. The reported stem cells for the treatment of cardiovascular diseases include embryonic stem cells, mesenchymal stem cells, endothelial progenitor cells, CD133$^+$ cells, cardiac stem cells, umbilical cord blood stem cells, adipose stem cells, and iPS cells.

The clinical efficacy of bone marrow mesenchymal stem cells in patients with heart failure was first confirmed by Hare and others. He treated 30 patients with ischemic cardiomyopathy with left ventricular dysfunction by intracardiac injection of autologous and allogeneic bone marrow mesenchymal stem cells at doses of 2×10^7, 1×10^8, and 2×10^8

cells. Within 30 days, one patient was hospitalized for heart failure. In the 6-minute walk test of autologous bone marrow mesenchymal stem cell transplantation, the Minnesota Living with Heart Failure Questionnaire (MLHFQ) showed that the score was significantly improved. Vrtovec and other researchers divided patients with non-ischemic dilated cardiomyopathy with left ventricular systolic dysfunction into the transplantation group (28 cases) and the control group (27 cases). The transplantation group was treated with intracoronary infusion of $CD34^+$ stem cells. At the one year follow-up, the total mortality rate of patients in the $CD34^+$ stem cell transplantation group was significantly reduced. Subsequently, the team studied the long-term efficacy of $CD34^+$ stem cells in patients with dilated cardiomyopathy. After 5 years of observation, the mortality rate of heart failure in patients receiving $CD34^+$ stem cell therapy was significantly lower than that in the control group. Two studies have shown that $CD34^+$ stem cells can significantly improve ventricular remodeling, exercise tolerance, and can affect long-term prognosis.

Some digestive system diseases can be treated with stem cells, including Crohn's disease and cirrhosis.

Crohn's disease is a kind of intestinal inflammatory disease with unknown causes that can occur anywhere in the gastrointestinal tract, but it mainly occurs in the terminal ileum and right colon. In 2012, the Korea Food and Drug Administration approved autologous adipose-derived mesenchymal stem cells for the treatment of complex Crohn's disease complicated with anal fistula, which was fortunate to be one of the seven approved stem cell drugs in the world at that time. In addition to mesenchymal stem cells, hematopoietic stem cell transplantation can change the natural process of Crohn's disease and is an important treatment option for patients with refractory Crohn's disease who cannot undergo surgery. In 2008, Cassinotti and others published a phase I and phase II clinical study of autologous bone marrow hematopoietic stem cell transplantation for the treatment of Crohn's disease in Milan, Italy. The study included four patients with Crohn's disease with similar clinical symptoms, all of whom failed in immunosuppression and anti-tumor necrosis factor therapy, and two of them also underwent multiple surgical procedures. Crohn's disease was active at the time of cell transplantation. Autologous peripheral blood stem cells were all re-infused without any immune selection. Three months after transplantation, all patients achieved remission of clinical symptoms, and endoscopic findings of two patients were also significantly improved. During the follow-up, the symptoms of three patients continued to be relieved. Due to the existence of a large number of patients with liver diseases in China, clinical research on the treatment of liver diseases by various stem cell transplantations has been carried out extensively. A new clinical study from China observed the efficacy of stem cell transplantation in the treatment of refractory ascites due to cirrhosis. Thirty nine patients with refractory ascites were divided into the following: Control group, 16 cases, routine treatment; treatment group, 23 cases, stem cell transplantations on the basis of routine treatment. The results showed that stem cell transplantations have a good short-term effect in the treatment of refractory ascites due to cirrhosis, which can be further promoted.

Based on the advantage of promoting angiogenesis, stem cell therapy can be applied to patients with lower limb ischemia. The strategy of transplantation of whole bone marrow cells, bone marrow mononuclear cells, and peripheral blood mononuclear cells is used to achieve successful angiogenesis. At present, the commonly used methods of transplantation include arterial and intramuscular approaches.

The studies of Lu and others have confirmed that compared with bone marrow mononuclear cells, bone marrow mesenchymal stem cells can improve lower limb ischemia, increase blood flow, and promote ulcer healing. It is suggested that bone marrow mesenchymal stem cells may be a more tolerable and effective method. Li and others used bone marrow mesenchymal stem cells to treat lower limb ischemia disease and found significant improvement in clinical manifestations such as resting pain and skin temperature.

More and more scientists are studying the use of stem cells for beauty and anti-aging. In aesthetic plastic surgery, some stem cells can directly replace the artificial prosthesis to beautify the body and facial contours of patients, and have good therapeutic effects on fat transplantation, removal of scars and wrinkles, and so on. The advantages of stem cells used in plastic and beauty are low absorption rate, high survival rate, and long-lasting effect. Therefore, in recent years, stem cells have a tendency of becoming a new force of filling agent instead of artificial prosthesis.

Some clinical trials have shown that stem cells have certain effects on cosmetic surgery and adipose tissue reconstruction. As soft tissue fillers, autologous adipose derived stem cells have both advantages and disadvantages in cosmetic surgery. Due to high absorption rate, low survival rate, many complications, and other reasons, its clinical application is limited. The adipose tissue vascularization and survival rate are enhanced by continuous improvement of fat acquisition technology. In 2001, Zuk and others found that besides the shaped preadipocytes, adipose tissue also contains a group of cells that have multi-directional differentiation potential, which are similar in nature to mesenchymal stem cells. This kind of cell not only has the ability to differentiate into bone, cartilage, fat, myocardium, nerve, and other tissues but also has the ability to promote wound healing, regeneration of damaged tissues, and reduce scar formation, which can produce good repair and cosmetic effects on the aging skin of the human body. Such cells are called adipose derived stem cells (ADSCs). By treating aging skin with autologous ADSCs, the thickness of skin was significantly increased, and the collagen content in dermis also showed a significant increase. Some scholars have studied the application of autologous ADSCs in augmentation mammoplasty, and the plastic effect is good. Park and other researchers directly injected ADSCs into the dermis of the patient's facial crow's feet. The results showed that the facial crow's feet became lighter and the skin texture became more delicate, which had obvious anti-aging effects. Studies have shown that autologous ADSCs can secrete a large number of growth factors, including epidermal growth factor, vascular endothelial growth factor, and fibroblast growth factor, which can promote collagen synthesis, increase blood vessels, and change the skin texture. Yoshimura and others invented cell-assisted lipotransfer (CAL) therapy. It involves injection transplantation with a mixture of autologous ADSCs and adipocytes, which can effectively improve the survival period of fat transplantation and provides a better method for fat transplantation. In fact, in the clinical application of stem cells, sometimes a combination of two different stem cells can be used. For example, the transplantation of hematopoietic stem cells can improve the therapeutic effect of diabetes in combination with mesenchymal stem cells. In addition, stem cells can also be used in the screening of new drugs, tissue engineering seed cells, tissue engineering treatment, tissue damage repair, and other fields.

In the future, as more and more stem cell drugs are approved for marketing, many difficult diseases that the existing drugs cannot treat or have poor therapeutic effects on will be effectively treated.

5. Immune Cells, Treatment of Tumors

The occurrence of tumors is related to immunity. What is immunity? It is a physiological function of the human body. The human body has an immune system consisting of immune organs (bone marrow, spleen, lymph nodes, tonsils, small intestine, lymph nodes, appendix, thymus, etc.), immune cells (lymphocytes, mononuclear phagocytic cells, neutrophils, basophils, eosinophils, mast cells, platelets, etc.), and immunologically active substances (antibody, lysozyme, complement, immunoglobulin, interferon, interleukin, tumor necrosis factor, etc.). By relying on the immune system, the human body can recognize its own substances and foreign substances, and reject, destroy, or eliminate foreign substances through immune responses. The foreign substances in the human body include the alien bacteria, viruses, and tumor cells produced by the human body itself. These foreign substances are also called antigens because they can induce the body's immune response. Under normal circumstances, the human body may also produce a small amount of tumor cells, which can be cleared by immune cells.

Since the human body has such a good defense mechanism, why do diseases still occur? This is because when the body's immune system is damaged or dysfunctional, its ability to resist foreign substances is greatly reduced. In this case, increasing the number and quality of human immune cells can enhance the ability of immune cells to kill tumors. Immune cell therapy is based on this principle. The blood is extracted from the body of tumor patients to improve the number and quality of immune cells and enhance the patient's ability to resist tumors through *in vitro* isolation, culture, amplification, activation, and other operation steps.

In recent years, the immune cells used in clinical research and application have mainly included DC cells (dendritic cells) [17], CIK cells (cytokine-induced killer cells), NK cells (nature killer cells), and CAR-T cells (chimeric antigen receptor T cells). Among them, DC cells were first discovered in the lymph nodes of mice by American scholar Steinman in 1973. They were named after their many dendritic or pseudopodiform processes at maturity. The only function known to date for DC cells is the antigen presentation and they are the most powerful antigen-presenting cells. An antigen is any substance that can induce immune response. The antigen-presenting cell is a type of cell that ingests and processes foreign antigens, and presents antigen information to T lymphocytes to induce immune response.

For a long time after the discovery of DC cells, due to the limitations of biomedical technology at the time, people could not cultivate more dendritic cells *in vitro*, and the price was expensive. As a result, the research on DC cells did not go further. In the 1990s, human beings made great progress in biomedical technology. DC cells can be cultured *in vitro*, and the research on DC cells has made a breakthrough. At the end of the 20th century, the United States took the lead in conducting DC cell immunotherapy for tumors in humans, and the results were encouraging. Subsequently, DC cells became the star of tumor biotherapy, and also became the research hotspot of scientists fighting cancer all over the world. In the 21st century, scientists have found that DC cells play an important role in the treatment of asthma and other diseases, and are clinically used for the biological treatment of various tumors.

On April 29, 2010, the US Food and Drug Administration approved Provenge, the first cancer treatment vaccine (developed by Dendreon Corporation) for the treatment of advanced prostate cancer, making it the first vaccine approved for cancer treatment in the United States. It has created a new era of cancer immunotherapy. The Provenge vaccine fights against malignant tumors using the patient's own immune system, which consists of neuronal DC cells of the tumor patient carrying the recombinant prostatic acid phosphatase antigen. GRNVAC1/GRNVAC2 of Geron Corporation of the United States is a telomerase cancer vaccine consisting of mature DC cells, human telomerase RNA, and a part of lysosome targeting signals. GRNVAC1 is injected through the patient's skin, and DC cells enter lymph nodes through skin and direct cytotoxic T cells to release telomerase to kill tumor cells. On September 6, 2013, the European Commission authorized the Provenge vaccine for the treatment of asymptomatic or mildly metastatic male adult prostate cancer. Academician Xuetao Cao, a famous immunologist in China, presided over "the somatic cell therapeutic vaccine, antigen pulsed human dendritic cells (APDC)". Sequentially, using APDC and combination chemotherapy has achieved remarkable efficacy in the treatment of advanced colorectal cancer in phase II clinical trial after nearly a decade of development, and has entered the phase III clinical trial. This is the first and only immune cell therapy technology officially approved by the China Food and Drug Administration.

CIK cells are a group of heterogeneous cells derived from human peripheral blood mononuclear cells (PBMCs) cultured *in vitro* with various cytokines (such as anti-CD3 monoclonal antibody, IL-2, and IFN-γ) for a period of time. It is a novel immunocompetent cell with strong proliferative ability, strong cytotoxic effect, and certain immune characteristics. Because of the two membrane protein molecules, CD3 and CD56, also known as NK cell like T lymphocytes, they possess the strong anti-tumor activity of T lymphocytes and the non-MHC restricted tumor killing ability of NK cells. CIK cells have the advantages of rapid proliferation, wide spectrum of tumor killing, and high tumoricidal activity. For patients with advanced tumors who have lost their opportunities of surgery or have relapsed and metastasized, they can quickly relieve the clinical symptoms, improve the quality of life, and prolong the survival period. Most patients treated with CIK cells, especially those after chemoradiotherapy, may have phenomena such as alleviated or disappeared gastrointestinal symptoms, shiny skin, faded black spots, disappeared varicose veins, stopped hair loss, and even hair growth or white hair blackening, and other "young" performances, and the significantly recovered mental state or physical strength.

The combination of DC cells and CIK cells has a synergistic anti-tumor effect. After DC cells were co-incubated with CIK cells, the expression of co-stimulatory molecules on the surface and the antigen presentation ability of DC cells were significantly increased, while the proliferation ability and cytotoxic activity of CIK cells were also enhanced, so the treatment with DC-CIK cells was more effective than the treatment with CIK cells alone. Coculture of tumor antigen-loaded DC cells with CIK cells can stimulate the production of tumor antigen-specific T cells. Such DC-CIK cell therapy has both specific and non-specific tumor-killing effects, and is more active than the CIK cells activated by the stimulation of DC cells without loaded tumor antigen, which is often used in clinic and scientific research. In adoptive immunotherapy of DC-CIK cells, the final effector cells are CIK cells activated by DC cells *in vitro*.

In recent years, the clinical studies of DC-CIK cell adoptive immuno-therapy have shown that it has certain effectiveness in the treatment of malignant melanoma, prostate cancer, kidney cancer, bladder cancer, ovarian cancer, colon cancer, rectal cancer, breast cancer, cervical cancer, lung cancer, laryngeal cancer, nasopharyngeal cancer, pancreatic cancer, liver cancer, stomach cancer, leukemia, etc.

NK cells are effector cells of the body's innate immune system, which are not only related to anti-tumor, anti-virus infection, and immune regu-lation but also involved in hypersensitivity and autoimmune diseases in some cases. Both bone marrow hematopoietic stem cells and early thymic lymphoid cells can develop and differentiate into NK precursor cells, and then develop into NK cells. The killing activity of NK cells is independent of antibodies and does not require specific antigen stimulation, and is called natural killing activity. NK cells are rich in cytoplasm and contain large azurophilic particles. The azurophilic granules are lysosomes containing acid phosphatase, myeloperoxidase, and various acidic hydro-lases, which can digest bacteria and foreign bodies phagocytized by cells, so the content of azurophilic granules is positively correlated with NK cell killing activity. After NK cells act on target cells, the killing effect appears early, and the killing effect can be seen in 1 hour *in vitro* and 4 hours *in vivo*.

The target cells of NK cells mainly include some tumor cells, virus-infected cells, some self-tissue cells (such as blood cells), and parasites. The anti-tumor mechanisms of NK cells mainly include the following: first, killing tumor cells by releasing cytotoxic particles; second, killing tumor cells by activating apoptosis of target cells with proteins synthe-sized on the cell surface; and third, killing target cells by the cytotoxicity of binding with tumor cell surface antibodies. In addition, NK cells can also secrete a variety of cytokines, such as TNF-α, TNF-β, IFN-γ, and so on, which can cooperate with its anti-tumor, and have a curative effect on lung cancer, breast cancer, colorectal cancer, liver cancer, leukemia, etc. NK cells are important immune factors for anti-tumor and anti-infection. They also participate in type II hypersensitivity and graft-versus-host reaction, and have the function of immune clearance and immune surveillance.

Cellular immunotherapy based on chimeric antigen receptor (CAR) is a new treatment mode of malignant tumors, which brings hope for the cure of some patients with advanced cancer. CAR-T cells are a novel targeted therapeutic approach for B-lymphocyte hematological malignancies by chimerism of T cell receptor gene and anti-CD19 antibody gene, transfected into T cells, amplified *in vitro* and re-infused to the patient. The genetically modified CAR-T cells have specific sites on the surface, which can recognize the CD19 antigen on the surface of B cells in B-lymphocyte hematological malignancies. Continuous stimulation of CD19 antigen can make CAR-T cells continuously proliferate and active and kill tumor cells effectively.

CAR-T cell therapy has undergone long-term research and development. In the late 1980s, the Israeli chemist and immunologist Zelig Eshhar developed the first CAR-T cell. In 1990, Zelig Eshhar came to the National Institutes of Health (NIH) on his annual leave to collaborate with Steven Rosenberg to study chimeric antigen receptors that target human melanoma. Zelig Eshhar and Steven Rosenberg constructed CAR-T cells in a modular design. CAR-T cell therapy once made a child who was only 6 years old and had nearly died of advanced leukemia miraculously break away from the bondage of death, and made more researchers see the dawn of hope, making CAR-T cell therapy become a hot area of tumor immunotherapy in recent years.

At present, CAR-T cells have made rapid progress in the treatment of acute and chronic lymphocytic leukemia, B-cell lymphoma, and the US Food and Drug Administration has given priority to the evaluation of CAR-T cell therapy. In November 2014, the US Food and Drug Administration granted "JCAR015" of the Juno Therapeutics, Inc. certification of "rare disease therapy". "KTE-C19" of the Kite Pharma, Inc. for the remission of non-Hodgkin's lymphoma has also been certified by the US Food and Drug Administration and the European Medicines Agency.

All of the above immunocyte therapies belong to the medical technology of category III. What is the medical technology of category III?

The medical technologies of category III refer to the medical technologies that need to be strictly controlled and managed by the national health administration under one of the following situations: first, involving major ethical issues; second, high risk; third, the safety and effectiveness need to be further verified by standardized clinical trials; fourth, using scarce resources; and fifth, other medical technologies requiring national special management.

The clinical effects of different immune cell therapies may vary greatly due to different patients and different diseases. At present, immunocyte therapy of tumors is usually used as an adjuvant treatment after radiotherapy, chemotherapy, and surgery. In the future, with the continuous development of technologies, immunocyte therapy may become one of the main methods of tumor treatment.

6. Ordinary Somatic Cells, Disease Therapy

Besides some stem cells and immune cells, are there any other cells applied in clinical therapy? The answer is yes. In fact, some ordinary somatic cells are also used in clinic. The reason why these cells are so-called ordinary somatic cells is that they do not have more basic research and clinical applications than stem cells and immune cells. These ordinary somatic cells mainly include chondrocytes, epidermal cells, fibroblasts, islet cells, hepatocytes, and corneal epithelial cells, all of which have completed cell differentiation, playing a specific structural role in specific tissue or organ, and performing certain functions. Their structural and functional effects are usually limited. It is not like stem cells that can be transformed into one or more other kinds of cells and have other structures and functions, nor do immune cells have important defense functions against diseases.

The transplantation of such cells still has important therapeutic value in clinic. In October 2009, the Committee for Medicinal Products for Human Use of European Medicines Agency (EMA) agreed with the positive opinion of the Committee for Advanced Therapy and recommended the marketing application of medical product Chondrocelecte suspension (containing 10,000 cells per microliter). ChondroCelect that comes from autogenous chondrocytes can be applied to repair single symptomatic cartilage injury of femoral condyle of adult knee joint. As an advanced medical product containing living cells, it is currently sold in Belgium, Holland, Luxembourg, Germany, Britain, Finland, Spain, and other countries.

MACI graft and Carticel, two products of autologous chondrocyte repair technology, developed by the Genzyme Co. in the United States, can replace injured knee cartilage. Carticel is the first cell therapy product approved by the US Food and Drug Administration, and also the first-generation product of autologous chondrocyte implantation (ACI) technology of the Genzyme Co., and MACI graft is mainly used by orthopedic surgeons to treat patients with clinically significant symptoms of articular cartilage injury. Both of them culture and transplant the patient's own chondrocytes to repair cartilage injury. Carticel autologous chondrocyte transplantation is mainly used to repair symptomatic femoral cartilage defects (medial, lateral, or trochlear) caused by acute or repeated trauma in patients with femoral condylar injury and poor response to previous arthroscopic or other surgical repair procedures (such as debridement, microfracture, drilling, and grinding arthroplasty). MACI graft is currently available in Europe, Asia, and Australia, while Carticel is available in the United States. In October 2007, the US Food and Drug Administration approved the launch of Epicel from the Genzyme Co. for the treatment of life-threatening severe burns. Epicel contains the patient's own epidermal cell component, which can provide permanent skin substitutes for burn patients. This is the first heterologous transplant system with living cells on the market in the United States, and later the company developed other artificial skin products containing living cells.

Dermagraft-TM is an artificial dermis made by Advanced Tissue Sciences Inc. It involves inoculating the fibroblasts obtained from the foreskin of newborns on the bioabsorbable polylactic acid (PLA) scaffold. After cultivating for 14 to 17 days, fibroblasts massively proliferate on the scaffold and secrete a variety of matrix proteins, such as collagen, fibronectin, and growth factors, and then the artificial dermis Dermagraft-TM composed of fibroblasts, extracellular matrix, and biodegradable biomaterials is formed. Its structure is more similar to natural dermis, which can reduce wound surface contraction and promote epidermal adhesion and basal membrane differentiation. Dermagraft-TM can be used not only for burn wound surface therapy but also for chronic skin ulcer wound surface therapy. In a randomized controlled clinical trial of 314 diabetic chronic foot ulcers in 35 medical centers in the United States, the safety and efficacy of Dermagraft-TM therapy was validated.

Dermagraft-TC is another artificial dermis produced by the Advanced Tissue Sciences Inc., which inoculates neonatal foreskin fibroblasts onto a membrane consisting of a layer of silica gel film and a nylon mesh attached to it. Dermagraft-TC is often used as a temporary dressing for burn wound surfaces. A multicenter study showed that the average burn area of 66 burn patients was 44%, and the acceptance rates of Dermagraft-TC and allogeneic skin after transplantation for 14 days were 94.7% and 93.1%, respectively. From the points of view of adhesion and empyema, there are no differences between the two groups. Dermagraft-TC is easy to remove, and it does not easily cause wound surface bleeding.

Apligraf produced by the Organogenesis Holdings, Inc. in the United States in 1998 is the most mature artificial skin with both epidermis and dermis. Apligraf is prepared by first inoculating neonatal foreskin fibroblasts into bovine collagen gel to form cell collagen gel, and then inoculating keratinocytes into the gel, and eventually culturing together. It has been approved by the US Food and Drug Administration (FDA) for repairing small-area surface wounds such as diabetic ulcers and venous ulcers. The results of the treatment of 208 patients with non-infectious neuropathic diabetic foot ulcers from 24 centers in the United States showed that 63 of 112 patients in the Apligraf treatment group were completely healed, while only 36 of 96 patients in the control group treated with wet gauze were healed; the average healing time of the former was 65 days, and the latter was 90 days. Clinical studies have also shown that Apligraf is more economical and effective than traditional methods in the treatment of venous ulcer. Apligraf can also be applied to treat epidermolysis bullosa, pyoderma gangrenosum, ulcerative sarcoidosis, and other diseases. Usually, metabolic diseases are attributed to certain cellular dysfunctions or defects, and cell transplantation is logically applied to the treatment of these diseases, such as islet cell transplantation [18]. Islet is a cell mass composed of tens to thousands of cells, generally round or oval in shape with different sizes. A few islets are irregular, such as half-moon shape and curved cylinder shape. According to the characteristics of staining and morphology, human islet cells are mainly divided into A (α) cells, B (β) cells, D cells, and PP cells.

A cells that account for about 20% of pancreatic islet cells can secrete glucagon and increase blood glucose; B cells that account for 60% to 70% of islet cells can secrete insulin and reduce blood glucose; D cells that account for 10% of islet cells can secrete growth hormone-inhibiting hormone (GHIH, also known as somatostain); and PP cells have only a small number and can secrete pancreatic polypeptide. Among them, B cells are functional cells for the treatment of diabetes, mainly distributed in the central part of the islets, arranged in a regular manner, generally round, uniform in size, more cytoplasm, filled with coarse insulin staining particles in the cytoplasm, and the nucleus is not stained. However, the outline of nucleus can be clearly observed, and it is mostly circular or elliptical. Therefore, B cells can be digested and isolated from islet tissue by collagenase and so on to treat diabetes.

The liver is made up of hepatocytes, and 50 hepatocytes form a hepatic lobule. It is estimated that the total number of hepatic lobules in human liver is about 500,000. The Hepatocyte is polygonal, at about 20–30 microns in diameter, and has 6–8 faces. The size of hepatocytes varies under different physiological conditions. For example, the hepatocytes enlarge during starvation. The hepatocytes have many functions, so it seems logical to transplant healthy hepatocytes for patients with hereditary metabolic defects.

Hepatocytes for transplantation therapy include the following:

(1) Xenogenic hepatocytes. At present, the most commonly used is pig liver, which can provide hepatocytes with similar structure and functions to human liver. Nishitai and others transplanted porcine hepatocytes into the spleen of immunodeficient mice, and found that the freshly isolated porcine hepatocytes have better activity and secretory function after transplantation than the cultured porcine hepatocytes after preservation at 4°C or cryopreservation in liquid nitrogen. It is suggested that the freshly isolated porcine hepatocytes are the first choice of xenogenic hepatocytes.

(2) Mature human hepatocytes. This is the most ideal source of hepatocytes.
(3) Fetal hepatocytes. Fetal hepatocytes are hepatocytes and their precursors isolated from the aborted fetal liver. They have the advantages of strong differentiation and proliferation abilities, weak immunogenicity, and better resistance to low-temperature storage injury.
(4) Immortalized hepatocyte line. Chinese scholars have successfully constructed immortalized human hepatocyte line HepLL by transfecting the recombinant plasmid SV40LT/pcDNA3.1 into normal human hepatocytes with liposome. The studies have shown that HepLL has the morphological characteristics and biological functions of normal human hepatocytes.

In 1976, Matas and others reported that the injection of hepatocytes from portal vein reduced plasma bilirubin levels in the Crigler-Najjar model rats. In 1992, human hepatocyte transplantation was successful in the first clinical trial. In 1993, Mito and others first reported the application of hepatocyte transplantation in the treatment of chronic severe hepatitis. In 1998, Fox and others reported the application of this method to treat type 1 Crigler-Najjar syndrome in children. This is a rare genetic disease, in which the symptoms of bilirubin encephalopathy such as muscle spasm, rigidity, convulsion, and opisthotonos are usually shown within 2 weeks of birth. After treatment, the bilirubin level in children was reduced by 60% after 18 months. In 1998, the US Food and Drug Administration's clause 6880 approved the transplantation of human hepatocytes *in vivo* as an effective treatment method for end-stage liver disease, and passed the US Food and Drug Administration certification that year. In recent years, a number of medical institutions in China have been carrying out hepatocyte transplantation therapies and have made some achievements.

Other ordinary somatic cells are also used for clinical treatment. For example, in February 2015, the European Medicines Agency approved the launch of Holoclar, developed by the Chiesi Farmaceutici S.p.A, Italy, for the treatment of moderate to severe limbal stem cell defects caused by physical or chemical burning in adult patients. The main component of Holoclar is corneal epithelial cells. With the deep understanding and skillful manipulation of the cell growth and development environment and regulatory mechanisms, more and more ordinary somatic cells will be used in clinic practices in the future, and the applied range of each type of human cell will be expanded.

The modern clinical trial research has shown that the certain types of cell mixed transplantation can improve the therapeutic effect. The transplantation of ordinary somatic cells and stem cells together will be an important development direction for future cell transplantation therapy, because the mature ordinary somatic cells can be used as inducers to induce the directional development and differentiation of stem cells, so that the stem cells can form tissues or organs with specific functions faster and better to enhance the therapeutic effect of stem cells and achieve the purpose of effectively treating diseases.

The 21st century is the era of cell transplantation therapy. In the next few decades, a large number of stem cell products will be approved for clinical application, while the transplantation therapy of immune cells, ordinary somatic cells, and mixed cells will also develop rapidly. Cell transplantation therapy will change our lives and make our lives better.

Appendix

1. Cell Transplantation Therapy Memorabilia

In 1667, the French doctor Jean-Baptiste Denis injected calf blood into a mentally ill patient, which is the first documented cell therapy.

In 1796, the British doctor Jenner Edward inoculated people with a vaccinia virus vaccine to prevent variola virus infection, which is the earliest biotherapy in the world.

In 1867, the German pathologist Cohnheim intravenously injected insoluble dye aniline into animals. As a result, the cells containing dyes, including inflammatory cells and fibroblasts associated with fiber synthesis, were found at the distal end of the animal's injury. From this, he inferred the presence of non-hematopoietic stem cells in the bone marrow.

In 1912, Kuettner, a German doctor, proposed that an organ should be cut into small pieces of tissue, dissolved in physiological saline, and injected into the body of a patient instead of using the whole organ for transplantation. Hence, Kuettner was a pioneer of cell therapy.

In 1956, Professor E. Donnall Thomas of the University of Washington, USA, completed the first bone marrow transplant in the world, which was also the first stem cell transplant in the world. E. Donnall Thomas thus became the founder of hematopoietic stem cell transplantation.

In 1981, Evan Kaufman and Martin isolated embryonic stem cells from a mouse blastocyst inner cell mass, established suitable culture conditions of embryonic stem cells *in vitro*, and bred them into stem cell lines.

In 1982, Grimm and others first reported that IL-2 was added to peripheral blood mononuclear cells for 4–6 days *in vitro* to induce non-specific killer cells, which are LAK cells.

In 1984, the Rosenberg research group was given approval by the US Food and Drug Administration for the first time to use IL-2 and LAK (lymphokine active killer cell) to treat 25 patients with renal cell carcinoma, melanoma, lung cancer, colon cancer, and other tumors with significant efficacy.

In 1930, Paul Niehans of Switzerland injected cells isolated from the embryonic organs of sheep into the human body for skin rejuvenation treatment. In the following year, a small tissue block cut from the bovine thyroid gland was dissolved in physiological saline to treat "hypothyroidism", which is known as "the father of cell therapy".

In 1967, Professor E. Donnall Thomas of the University of Washington, USA, published an important paper on stem cell research in *The New England Journal of Medicine*. This paper described in detail the hematopoietic principle of stem cells in bone marrow, the process of bone marrow transplantation, and the role of stem cells in patients with hematopoietic dysfunction. This paper presented a broad prospect for the treatment of hereditary diseases such as leukemia, aplastic anemia, thalassemia, and immune system diseases. Since then, stem cell research has attracted the attention of biologists and medical scientists in various countries, and stem cell transplantation has been rapidly carried out in countries around the world.

In 1973, American scholars Steinman and Cohn discovered dendritic cells (DCs) in the spleen tissue of mice, which were so named because of their dendritic or pseudopodia-like processes.

In 1985, Rosenberg was the first to report that interleukin-2 (IL-2) and lymphokine-activated killer cells (LAK cells) were effective in the treatment of advanced tumors.

In 1987, Peterson used autologous chondrocyte implantation (ACI) to treat patients with articular cartilage defects. This is the first time that cell engineering technology has been used in the treatment of osteoarthrosis and has become a relatively mature technique for the treatment of articular cartilage defects.

In 1988, the first batch of tissue repair cell therapy entered the market, and they were wound healing products for the treatment of severe burns.

In 1989, a scientist in the United States discovered neural stem cells in brain tissue.

In 1990, E. Donnall Thomas won the Nobel Prize in Physiology or Medicine for his pioneering work in stem cell transplantation.

In 1990, Scharp and others reported the success of the first human allogeneic islet cell transplantation for type 1 diabetes.

In 1992, the first clinical trial of human hepatocytes transplantation was successful.

In 1994, peripheral blood mononuclear cells were induced to differentiate into CIK cells by Schmidt-wolf. CIK cells have a combination of strong tumoricidal activity of T lymphocytes and non-MHC restriction of NK cells, and are also called NK cell like T lymphocytes. It has been proved with significant effect that the CIK cells with high killing activity and the DCs with strong tumor antigen presenting ability can be cocultured to treat malignant tumors.

In 1997, Asahara and coworkers first described endothelial progenitor cells. They isolated a group of cells from peripheral blood mononuclear cells that could differentiate into endothelial cells under suitable conditions *in vitro*. The cell surface specifically expressed hematopoietic stem cell markers CD133, CD34, and endothelial cell marker VEGFR-2.

In 1998, two American scientists, James A. Thomson and John D. Gearhart, established human embryonic pluripotent stem cell lines.

In 1999, American scientists reported in the *Proceedings of the National Academy of Sciences of the United States of America* (PNAS) that adult stem cells from mouse muscle tissue can be differentiated laterally into blood cells. Subsequently, scientists from all over the world confirmed that adult stem cells, including human adult stem cells, have plasticity.

In 1999, stem cell research was selected by the American journal of *Science* as one of the 10 most important research fields in the 21st century, and ranked the first, ahead of the great engineering "human genome sequencing".

In 2000, Japan launched the "Millennium Century Project", which took stem cell engineering as the core technology of regenerative medicine and as one of the four major priorities, and the amount of investment in the first year was 10.8 billion yen.

In 2000, stem cell research was once again selected as one of the first of the top ten scientific and technological achievements of that year by the American journal of *Science*.

In 2001, the upper house of the British Parliament passed a bill with 212 votes in favor and 92 votes against, allowing scientists to clone human early embryos and use them for medical research. After cloning early human embryos using human cells, embryonic stem cells that are not fully developed can be extracted therefrom.

In 2001, some French scholars jointly submitted an investigation report to the French Minister of Scientific Research, calling on the government to strengthen the support for stem cell research.

In 2001, American scientists reported in the journal of *Tissue Engineering* that the adipose tissue extracted from hips and thighs contained a large number of stem cell-like cells that could develop into healthy cartilage and muscle.

In 2001, a British company announced the development of an umbilical cord blood stem cell storage service for newborn babies. When the parents spend 600 pounds, the newborn baby's umbilical cord blood can be collected and the stem cells can be isolated from it and stored in liquid nitrogen for at least 20 years.

In 2001, China completed the transplantation of human neural stem cells and corneal stem cells.

In 2001, the Tianjin Umbilical Cord Blood Hematopoietic Stem Cell Bank was officially put into operation.

In 2008, China's first stem cell hospital was established in Tianjin. It was combined with the Tianjin umbilical cord blood hematopoietic stem cell bank and the Tianjin mesenchymal stem cell bank to form a relatively complete stem cell engineering system integrating the research, development, storage, and application of stem cell products.

In 2009, the US Food and Drug Administration (FDA) approved the clinical trial of using embryonic stem cells for the treatment of paraplegia patients for the first time. Since then, stem cell research has entered the clinical stage through the fast track. As of January 2009, 20 clinical trials have been registered on the National Institutes of Health's website of "clinicaltrials.gov", and early results are encouraging.

In 2009, the Management Measures for Clinical Application of Medical Technologies promulgated by the Ministry of Health of China provided the guidance and basis for the strict and orderly development of cell biotherapy, and also ensured the safety and standardization of biotherapy.

In 2010, human embryonic stem cells were injected into the human body for the first time for clinical trial of disease treatment.

In 2010, the Australian Therapeutic Goods Administration (TGA) approved the autologous mesenchymal precursor cells from Mesoblast Ltd for the treatment of bone injury.

In 2011, the Korea Food and Drug Administration (KFDA) approved FCB-Pharmicell Co., Ltd's Hearticellgram-AMI (also called "Cellgram-AMI"), the autologous bone marrow mesenchymal stem cells, for the treatment of acute myocardial infarction.

In 2011, the US Food and Drug Administration approved allogeneic hematopoietic stem cell transplantation by using Hemcord, the cord blood hematopoietic progenitor cells from the New York Blood Center, for the treatment of hereditary or acquired hematopoietic diseases.

In 2012, the US Food and Drug Administration approved ELELYSO™ (taliglucerase alfa), produced by genetically engineered plant cells for the treatment of type 1 Gaucher's disease.

In 2012, the European Commission (EC) approved Glybera, the first gene therapy drug in the western world, marking a milestone in the novel medical technologies for repairing genetic defects. The drug is used to treat an extremely rare genetic disease, lipoprotein lipase deficiency (LPLD), which will be a major driver of gene therapy.

In 2012, the Korea Food and Drug Administration approved the Medipost Co., Ltd's cord blood-derived mesenchymal stem cell drug Cartistem for the treatment of degenerative arthritis and knee articular cartilage injury.

In 2012, the Korea Food and Drug Administration approved the Anterogen Co., Ltd's autologous adipose-derived mesenchymal stem cell drug Cuepistem for the treatment of complex Crohn's disease complicated with anal fistula.

In 2012, Health Canada (HC) approved the application of allogenic bone marrow mesenchymal stem cells (Prochymal) from the Osiris Therapeutics, Inc., USA to treat acute graft-versus-host disease (GVHD) in children.

In 2013, the US *Science* journal listed tumor immunotherapy as the first of the top ten scientific breakthroughs of the year, and determined the important position and development prospects of biological immunotherapy in the future comprehensive treatment of tumors.

In 2013, the first Chinese hamster ovary (CHO) cell line genome map was completed.

In February 2014, Japan established a pharmaceutical manufacturer called SIREGE (means sight regeneration) using induced pluripotent stem cells (iPS cells) as drugs, and it is committed to using iPS cells to treat age-related macular degenerative eye disease.

In February 2015, the European Medicines Agency approved the clinical application of human autologous corneal epithelial cell drug Holoclar containing stem cells from Italy's Chiesi Farmaceutici S.p.A for the treatment of moderate to severe limbal stem cell deficiency (LSCD) caused by physical or chemical burning in adult patients.

In June 2015, the European Medicines Agency approved Stempeusel, a stem cell drug that was obtained from the first Chinese patent by Stempeitics Institute, for the treatment of thromboarteritis obliterans.

In September 2015, Japan's Ministry of Health, Labour, and Welfare (MHLW) approved temcell, a bone marrow mesenchymal stem cell drug applied by JCR Pharmaceuticals Co., Ltd in Japan, for the treatment of acute graft-versus-host disease (GVHD), one of the serious complications after hematopoietic stem cell transplantation.

In December 2016, the US Food and Drug Administration approved a tissue engineering product Maci formed by autologous chondrocyte culture for the treatment of knee articular cartilage injury, which was developed by the Vericel Co. in the United States.

In August 2017, the US Food and Drug Administration approved Kymrah^{TM} (tisagenlecleucel) chimeric antigen receptor T cells (CAR-T) intravenous infusion suspension that was developed by the Novartis Pharma Co. for the treatment of patients aged 3–25 years with acute lymphoblastic leukemia.

In October 2017, the US Food and Drug Administration approved Yescarta (axicabtagene ciloleucel) chimeric antigen receptor T cells (CAR-T) intravenous infusion suspension that was developed by the Kite Pharma Inc. for adult patients with relapsed or refractory large B-cell lymphoma.

On January 12, 2018, the Nanjing Gulou Hospital ushered the first healthy baby born in the clinical research of stem cell treatment of premature ovarian failure in China. It was also the first healthy baby born in the clinical research of stem cell composite collagen scaffold material for the treatment of premature ovarian failure in the world, and was a major breakthrough in stem cell tissue engineering and regenerative medicine.

In March 2018, the European Commission approved allogeneic adipose derived mesenchymal stem cell drug Alofisel that was developed by Belgium's TiGenix Biopharmaceutical Company and Takeda Pharmaceutical Company Limited of Japan for the treatment of adult patients with non-active / mildly active luminal Crohn's disease complicated with complex perianal fistula.

In August 2020, the Drug Controller General of India (DCGI) approved an adult allogeneic bone marrow mesenchymal stem cell drug named Stempeucel to launch in India for the treatment of critical limb ischemia (CLI) due to Buerger's Disease and Atherosclerotic Peripheral Arterial Disease. The product was developed by the partner of Cipla Limited, Stempeutics Research Pvt. Ltd.

2. Animal Cloning Events

In 1938, Hans Spemann first proposed the idea of cloning animals using cell nucleus transfer technology, and called it a "fantastical" experiment. Later, the famous cloned sheep Dolly was cloned using the same idea.

In 1952, American scientists Robert Briggs and Thomas J. King used a tadpole cell to create an exact replica of the original tadpole. The little tadpole changed the history of biotechnology and became the first cloned animal in the world.

In the spring of 1979, scientists from the Wuhan Institute of Aquatic Biology of the Chinese Academy of Sciences cloned a crucian successfully, which was an embryonic cell clone.

In 1989, the United States obtained an embryonic cloned pig.

In 1991, China obtained an embryonic cloned goat.

In July 1996, the world's first adult somatic cell cloned sheep, Dolly, was born at the Roslin Institute in Edinburgh, a city in the north of England. It was the first demonstration that animal somatic cells are the same genetic totipotency as plant cells, breaking the traditional scientific concept and making a sensation in the world.

In July 1997, the Roslin Institute in Edinburgh successfully bred a cloned sheep Polly carrying a human α-antitrypsin gene. The human α-antitrypsin in sheep milk is a specific drug for treatment of chronic emphysema, congenital pulmonary fibrosis cyst, and other diseases.

In 1998, scientists from the University of Hawaii in the United States cloned more than 50 mice using adult mouse cells and then bred three generations of experimental mice with identical genetic characteristics. At the same time, several other private research institutions successfully cloned calves in different ways. Among them, it was the most remarkable that Japanese scientists used cells from an adult cow to produce eight calves with identical genetic characteristics, and the success rate was as high as 80%. Since then, large-scale cloning has begun.

On October 15, 1999, the Institute of Developmental Biology of the Chinese Academy of Sciences and Yangzhou University jointly studied and successfully cloned a transgenic goat in Yangzhou, Jiangsu Province. This is the first somatic cell transgenic clone goat in China, which can be used for the production of the rare drug human lactoferrin (hLF). The achievement of Somatic Cell Transgenic Clone Goat was ranked at the top out of the top ten news of China's basic research in 1999.

On December 24, 1999, the Biotechnology Research Center of Shandong Academy of Agricultural Sciences and Hebei Agricultural University jointly studied the key problems and successfully cloned two small white rabbits, which were named as "Luxing" and "Luyue" by the experts and grew well.

On April 13, 2000, researchers in Oregon, USA, cloned a monkey using a totally different embryonic division method than that used for the cloned sheep Dolly. The scientists divided an early embryo containing only 8 cells into 4 parts, and then cultivated them to produce new embryos. The only survivor was the short-tailed monkey Tetra. Unlike Dolly, Tetra had a mother and a father, but it was just one of the artificial quadruplets.

On January 23, 2000, the first re-cloned cattle in the world was born in Japan, and the somatic cell used for the re-cloning was obtained from a cloned cattle.

On March 14, 2000, the British PPL Biotech Co. announced to the press that, for the first time, by using technology similar to the clone sheep Dolly, they had successfully cloned five piglets that could be used for human organ transplantation. The piglets were named Millie (meaning the new millennium), Christiaan (commemorating the surgeon Christiaan Barnard who performed the first human heart transplantation operation in 1967), Alexis and Carrel (named after Alexis Carrel, a Nobel laureate and pioneer of organ transplantation), and Dotcom (i.e., .com, a worldwide popular Internet domain name at that time).

On June 16, 2000, an adult somatic cell cloned goat named Yuanyuan was born in Northwest A&F University. On June 22 of the same year, the university successfully cloned another goat, Yangyang.

On June 28, 2000, British scientists announced that they had mastered a new technology that could carry out precise genetic modification of large mammals. However, large-scale production of these animals still needed cloning.

In December 2000, the well-known Roslin Institute and a biotechnology company in America cooperated for more than two years to successfully cultivate transgenic cloned chickens. One of them laid eggs that could be extracted to get a new anti-cancer drug.

In 2000, scientists successfully cloned a pig, which is one of the most difficult animals to clone. The biotechnology company that helped cultivate Dolly the sheep announced the successful cloning of five piglets. The company announced that the cloned pigs would eventually become a processing factory for human organ transplantation. It also cloned the white-legged bison, which is of great significance for saving endangered animals. These important advances and the research of embryonic stem cells were rated as the fifth of the world's top ten scientific achievements by the American journal *Science*.

In 2001, scientists from America and Italy jointly launched the work of human cloning. In November 2001, American scientists announced the first successful cloning of human embryos at an early stage, claiming that their goal was to customize human cells for transplants that would not induce rejection.

On January 8, 2001, an Indian bison named Noah was successfully cloned in Iowa, USA. Only 12 hours after birth, the calf could walk without help.

On August 8, 2001, the somatic cell cloned goat "Yangyang" bred by the Northwest A&F University was naturally mated with the embryonic cell cloned goat "Shuaishuai" and gave birth to a pair of male–female goat twins.

In the first ten-day period of March 2002, China's first batch of self-contained adult somatic cell cloned cattle was successfully born in Cao County, Shandong Province, China. On the 20th of the same month, the Administrative Measures on the Identification of Agricultural Genetically Modified Organisms promulgated by the Chinese government began to be implemented.

In February 2004, the fourth generation of somatic cell cloned goat "Xiaoxiao" was born, which was cultivated by Professor Yong Zhang of Northwest A&F University.

On August 11, 2004, the United Kingdom issued the world's first human embryo cloning license, which was valid for only one year. The embryo had to be destroyed after 14 days. It is still illegal to breed cloned babies. Its purpose is to increase human understanding of the development of its own embryos, increase human understanding of high-risk diseases, and promote human research in the treatment of high-risk diseases.

In 2005, South Korean scientists used stem cell transplantation technology to breed the world's first cloned dog and named the cloned dog Snoopy.

On February 18, 2005, the 59th session of the UN General Assembly Law Committee adopted a political declaration in the form of a resolution with 71 votes in favor, 35 votes against, and 43 vote abstentions, requiring countries to ban any form of cloning human beings that violates human dignity.

On May 25, 2007, the world's first human–animal hybrid sheep was born from the work of a research group led by Professor Esmail Zanjani of the University of Nevada. This hybrid sheep containing 15% of human cells took the research team seven years. The purpose of this study was to solve the problem of transplantable organ shortage in the medical field by implanting human stem cells into animals and cultivating various organs suitable for transplantation.

On January 15, 2008, the US Food and Drug Administration announced that it approved the sale of dairy and meat products of cloned animals, and claimed that foods of this origin are safe.

On September 1, 2015, Russian media reported that the director of the Russian Mammoth Museum, Semyon Grigoriev, said that the first laboratory for cloning extinct animals in Russia began to work in Yakutsk. Their goal is to find living cells needed for cloning so that the mammoth can "regenerate."

In December 2015, China Boya Stem Cell Group Co., Ltd. cooperated with South Korean Sooam Biotech Research Foundation to build the world's largest cloning factory in Tianjin, China. It plans to produce 1 million cloned cattle each year, and will clone dogs and even endangered species. Xiaochun Xu, the chairman of the company, said that the company was able to achieve human cloning, but worried that the public would overreact, and would rather that the plan be postponed.

In November 2018, the international well-known academic journal *Cell Stem Cell* reported that scientists from the Institute of Zoology of Chinese Academy of Sciences had successfully bred double father mice with hybrid cells and clone technology. The hybrid cells were formed with the haploid embryonic stem cells, sperms, and enucleated oocytes by means of imprinted gene deletion and nuclear cytoplasmic hybridization technologies. The haploid embryonic stem cells were constructed by sperms. Imprinted genes only express the homologous genes of one parent in offspring.

References

1. Onodera, T. and Sakudo, A. Introduction toc progress in advanced research on prions. *Curr Issues Mol Biol.*, 2020, 36:63–66.
2. Riek, R., Hornemann, S., Wider, G., Billeter, M., Glockshuber, R., and Wüthrich, K. NMR structure of the mouse prion protein domain PrP (121–231). *Nature*, 1996, 382:180–182.
3. Hongxia, J., Zhangliu, C., Xuming, De., Zhenling, Z., Liancheng, L., Fengxia, S., Yumei, H., and Xiaohuan, Z. Ultrastructural comparison on sensitive and resistance strains of *Mycoplasma gallisepticum*. *Chinese J. Preve. Veterin. Medic.*, 2004, 26(6):439–442.
4. Blount, Z.D. The unexhausted potential of *E. coli*. *Elife*. 2015, 4:e05826.
5. Jin, Z., Jiangzhou, L., Xueli, Z., Meichao, W., Ran, Z., Jie, Z., Xiaotong, J., and Yilong, H. Establishment of anther culture technology system in blueberry variety 'Summit'. *Mol. Plant Breed.*, 2019, 17(20):6756–6761.
6. Danqing, T., Yaying, G., Xiaoyun, P., Liang, J., Yuan, Z., Qiang, Z., and Xiao, W. Anther culture and haploid identification of *Anthurium andraeanum. Mol. Plant Breed.*, 2020, 1:7.
7. Zaulyanov, L. and Kirsner, R.S. A review of a bi-layered living cell treatment (Apligraf) in the treatment of venous leg ulcers and diabetic foot ulcers. *Clin. Interv. Aging.*, 2007, 2(1):93–98.
8. Bhatia, S.N., Underhill, G.H., Zaret, K.S., and Fox, I.J. Cell and tissue engineering for liver disease. *Sci Transl. Med.*, 2014, 6(245):245sr2.

9. Keene, C.D., Cudaback, E., Li, X., Montine, K.S., and Montine, T.J. Apolipoprotein E isoforms and regulation of the innate immune response in brain of patients with Alzheimer's disease. *Curr. Opin. Neurobiol.*, 2011, 21(6):920–928.

10. Hwang, D. and Rader, C. Site-Specific antibody-drug conjugates in triple variable domain fab format. *Biomolecules*, 2020, 10(5):764.

11. Wilmut, I. and Taylor, J. Cloning after Dolly. *Cell Reprogram*, 2018, 20(1):1–3.

12. Mikkers, H.M., Freund, C., Mummery, C.L., and Hoeben, R.C. Cell replacement therapies: Is it time to reprogram? *Hum. Gene. Ther.*, 2014, 25(10):866–874.

13. Strom, C.M., Bonilla-Guererro, R., Zhang, K., Doody, K.J., Tourgeman, D., Alvero, R., Cedars, M.I., Crossley, B., Pandian, R., Sharma, R., Neidich, J., and Salazar, D. The sensitivity and specificity of hyperglycosylated hCG (hhCG) levels to reliably diagnose clinical IVF pregnancies at 6 days following embryo transfer. *J. Assist. Reprod. Genet.*, 2012, 29(7):609–614.

14. Suzuki, M. In vitro fertilization in Japan — early days of in vitro fertilization and embryo transfer and future prospects for assisted reproductive technology. *Proc. Jpn. Acad. Ser. B Phys. Biol. Sci.*, 2014, (5):184–201.

15. Atkins, H.L. and Freedman, M.S. Hematopoietic stem cell therapy for multiple sclerosis: Top 10 lessons learned. *Neurotherapeutics*, 2013, 10(1):68–76.

16. Kim, C., Lee, H.C., and Sung, J.J. Amyotrophic lateral sclerosis — cell based therapy and novel therapeutic development. *Exp. Neurobiol.*, 2014, 23(3):207–214.

17. Hubo, M., Trinschek, B., Kryczanowsky, F., Tuettenberg, A., Steinbrink, K., and Jonuleit, H. Costimulatory molecules on immunogenic versus tolerogenic human dendritic cells. *Front Immunol.*, 2013, 4:82.

18. Brian, F. Islet cell transplantation. *Semi. Inter. Radio.*, 2006, 23(3): 295–297.

Postscript

In June 2017, the book *Cells and Stem Cells: The Myth of Life Sciences* was published for the first time. Since then, several years have passed, and the field of cells and stem cells has developed rapidly, with much new progress and many research achievements made. In the process of writing this English version, the contents have been carefully revised. Many out-of-date words and images have been deleted, the latest achievements and research advances have been added correspondingly, and some references have been listed.

Index